Plumbing and Heating

William J. Hornung

Assistant Professor
California State Polytechnic University
Pomona, California

Prentice-Hall, Inc., Englewood Cliffs, New Jersey 07632

Library of Congress Cataloging in Publication Data

Hornung, William J.
Plumbing and heating.

Includes index.
1. Plumbing. 2. Heating. I. Title.
TH6124.H62 696'.1 81-10729
ISBN 0-13-683920-7 AACR2

Editorial/production supervision
and interior design by Lori Opre
Cover design by Mario Piazza
Manufacturing buyer: Joyce Levatino

Printed in the United States of America

10 9 8 7 6 5 4 3 2 1

ISBN 0-13-683920-7

Prentice-Hall International, Inc., *London*
Prentice-Hall of Australia Pty. Limited, *Sydney*
Prentice-Hall of Canada, Ltd., *Toronto*
Prentice-Hall of India Private Limited, *New Delhi*
Prentice-Hall of Japan, Inc., *Tokyo*
Prentice-Hall of Southeast Asia Pte. Ltd., *Singapore*
Whitehall Books Limited, *Wellington, New Zealand*

Contents

CHAPTER 4 LAWN SPRINKLER SYSTEMS 75

CHAPTER 5 PLUMBER'S TOOLS 88

CHAPTER 6 PLUMBING REPAIRS 105

CHAPTER 7 HEAT LOSS: HOW TO COMPUTE IT 119

CHAPTER 8 HEATING SYSTEMS 139

Preface

This book is for the beginning student of plumbing and heating whose aim it is to get into a marketable trade as well as for the homeowner who desires to make some of his or her own plumbing and heating repairs.

Both are provided with the basic principles of the house plumbing system; that is, the water distribution system, the sanitary drainage system, and the storm water drainage system.

A section on how to install your own lawn or garden sprinkler system for the do-it-yourselfer is included as well as a chapter on proper use of the common and essential plumber's tools. Fixing faucet leaks, servicing aerators on faucet spouts, replacing valve seats, installing a new lavatory, and setting and placing a new water closet are but some of the repairs and installations that are illustrated and discussed.

A chapter is devoted to home heat loss and a simplified method of how to compute it is followed by a review of the various types of heating systems in common use, together with some of the automatic equipment necessary for the proper functioning of the systems.

The written material is profusely illustrated to provide for a keener understanding of the subject matter. A series of review questions given at the end of each chapter challenges the reader and tests his or her comprehension of the subject.

Wherever English measurements appear in the text, their metric equivalents are generally used as well.

A chapter of review questions and answers containing a number of problems on plumbing and heating for research and supplementary study material concludes the text.

WILLIAM J. HORNUNG

Mission Viejo, California

1 Water Distribution System

1-1 INTRODUCTION

The fresh water in a water distribution system comes from a public works. It is first purified and then sent under pressure through pipes to the water service line that runs directly into houses, or other types of buildings at pressures of 55, 60, or 75 pounds per square inch (psi) and ranging as high as 100 psi [689.5 kilopascals (kPa)], Fig. 1-1.

1-2 MEASURING WATER PRESSURE

A water pressure gauge can be obtained from your plumbing supplier. Screw the gauge on to one of the sill cocks located outside of the house, and then turn on the faucet making certain no other faucets in other parts of the house are open. The pressure gauge shown in Fig. 1-2, registers 90 psi (620.5 kPa), a typical reading taken from a small newly constructed residential house.

Understanding, however, that high water pressures in a water system can eventually damage faucets and fittings. Water hammer, a banging noise caused by vibrating pipes, can occur when a valve abruptly stops the pressurized flow of incoming water. Water hammer

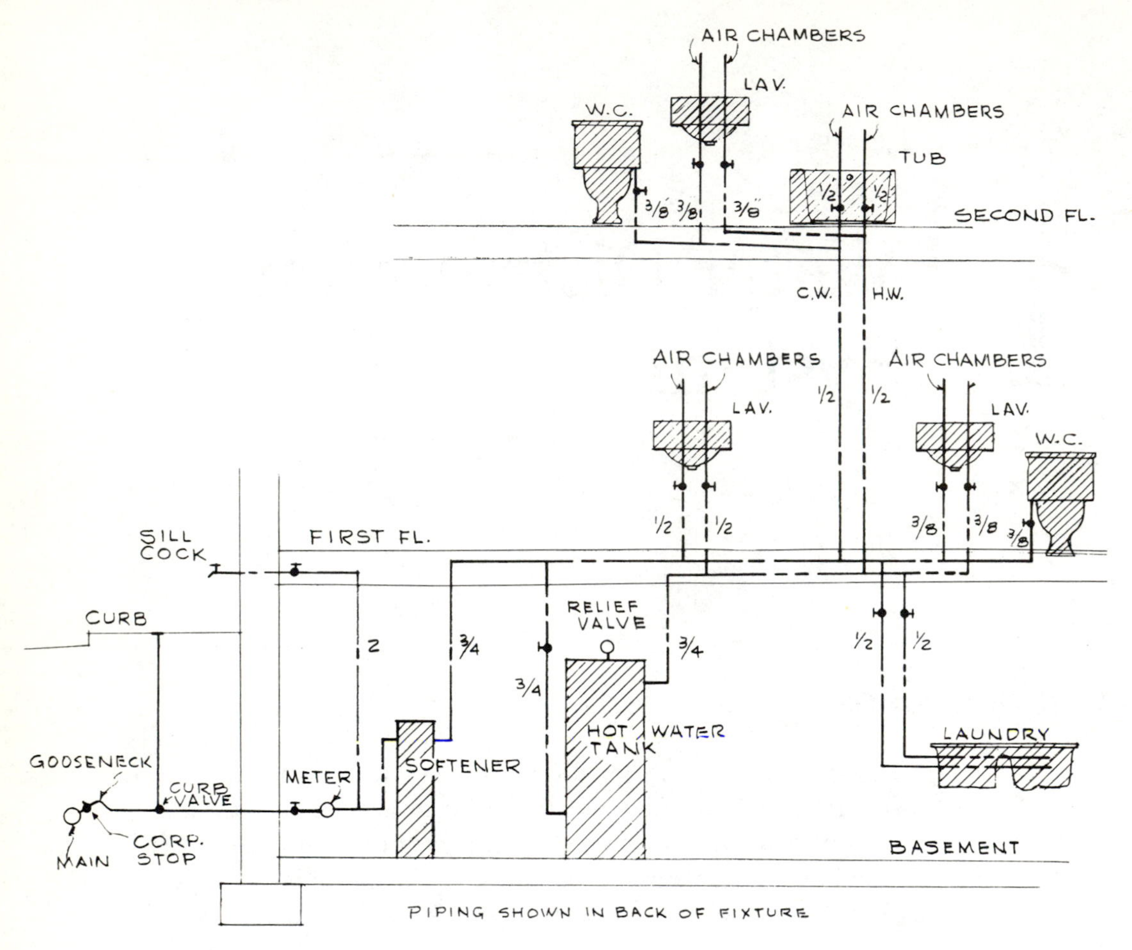

Fig. 1-1 Water distribution system.

can even occur if a high speed wall of liquid, as from a starting pump, hits a sudden change of direction in the piping such as an elbow. This problem can be eliminated by installing a ready-made shock absorber, or by assembling an air chamber from standard pipe and fittings, Fig. 1-3. By installing a pressure-reducing valve near the point where the water enters the house (Fig. 1-4), the line pressure of 90 psi (620.5 kPa) can be brought down to a reasonable 60, 55, or even 40 psi (413.7, 379.2, or 275.8 kPa).

To lower the pressure on existing installations, turn the adjusting nut on the pressure-reducing valve located just below the sill cock on the water service line clockwise. Turn the nut until the gauge shows the pressure desired.

Fig. 1-2 Water pressure gauge.

Fig. 1-3 Shock absorbers.

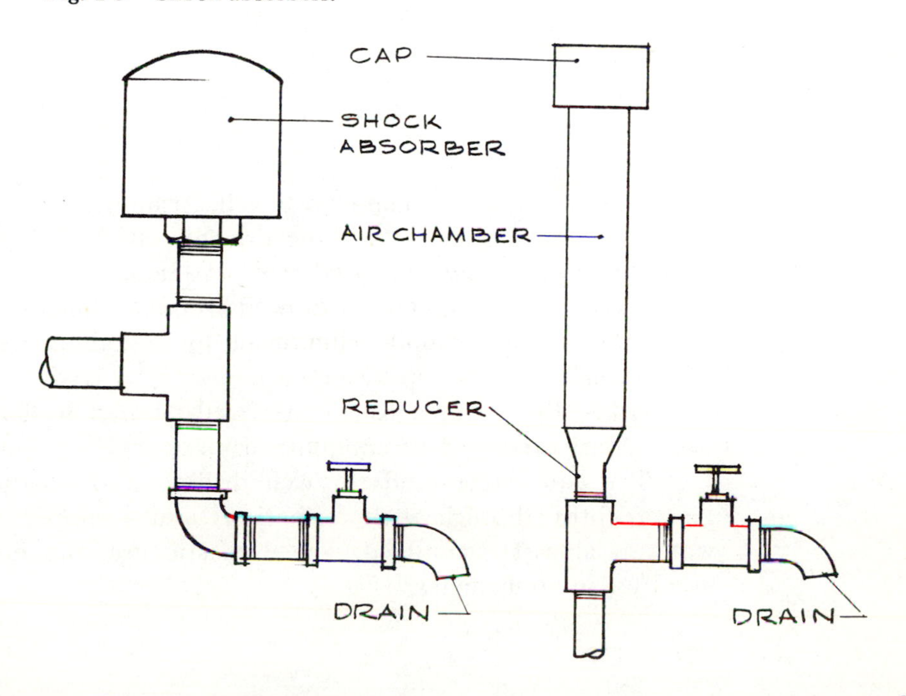

1. Pressure reducing valve
2. Sill cock
3. Gate valve

Fig. 1-4 Pressure reducing valve on water service from street.

1-3 WELL WATER

Fresh water may also come from a well. Although there are a variety of well types, the simplest is the driven well. This type of well is made by sledge hammering a pointed pipe (Fig. 1-5) into the ground. The pointed end has a screen to filter out sediment. The pipe is driven below the ground water level so that the water can enter the pipe and be drawn up by a pump.

Other types of wells are dug wells, bored wells, and drilled wells (considered the best and most often used).

The equipment used for well drilling is able to penetrate to great depths through rock formations and a supply of adequate water is almost guaranteed. For domestic use, the drilled well is usually 6 in. in diameter.

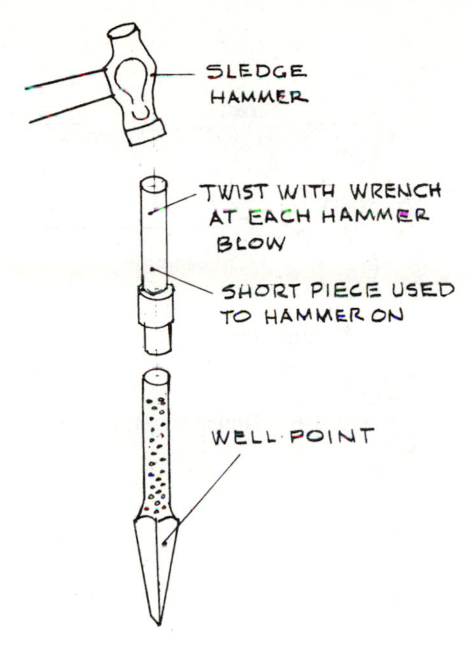

Fig. 1-5 Well point.

1-4 HOUSE SERVICE

The water distribution system supplies water to the fixtures first passing through the house service, then the house main supply lines, and finally to the fixture supply risers, branches and into the fixtures and appliances.

The house water service shown in Fig. 1-6, is that part of the water distribution system that runs from the street water main to the water meter that is usually inside the house. In warmer climates, such as southern California and Florida, the water meter is placed 1 ft. underground in the front of the house near the street, as shown in Fig. 1-7. A concrete or metal plate can be lifted off for a full view of the meter.

The *corporation stop* is a valve installed in the house service at the street water main. It can be turned on or off by the utility that supplys the water or the city, town, or village water department. If a building is to be demolished, the corporation stop is turned off.

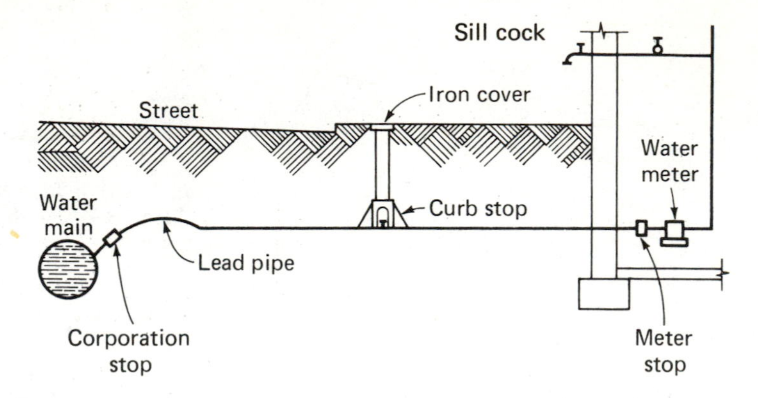

Fig. 1-6 House water service. Water meter inside of house.

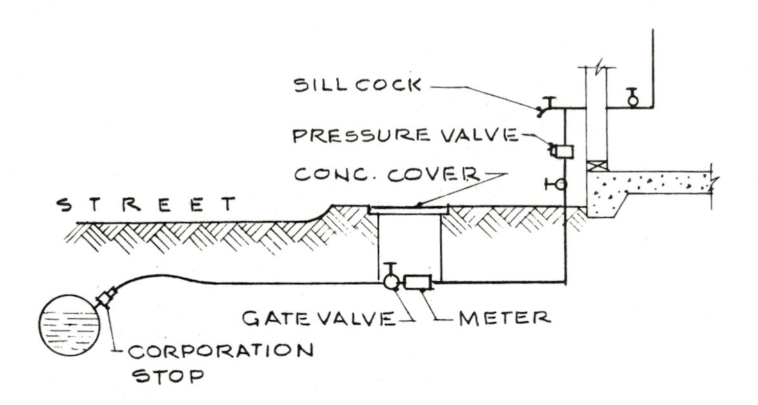

Fig. 1-7 House water service. Water meter outside of house.

The *curb stop* is another valve on the house service line. It is installed between the curb and the sidewalk. To get at the valve a cast iron stop box is brought up to ground level. The stop box is equipped with an iron cover that can be lifted off. A long rodlike wrench is used to turn the valve on and off. The curb stop is important since the water can easily be shut off in the event something happens to the piping system inside the house, such as a broken pipe, or when the building is not used during the winter.

The *water meter* measures the cubic feet or gallons of water coming through the house service. The meter is installed by, and is the property of the utility, city, town, or village. The meter is read regularly and the owner of the building is charged for the water used. One cubic foot (cu ft) of water is equal to 7½ gallons (gal) [28.4 litres (*l*)]. The meter can also be used for finding the

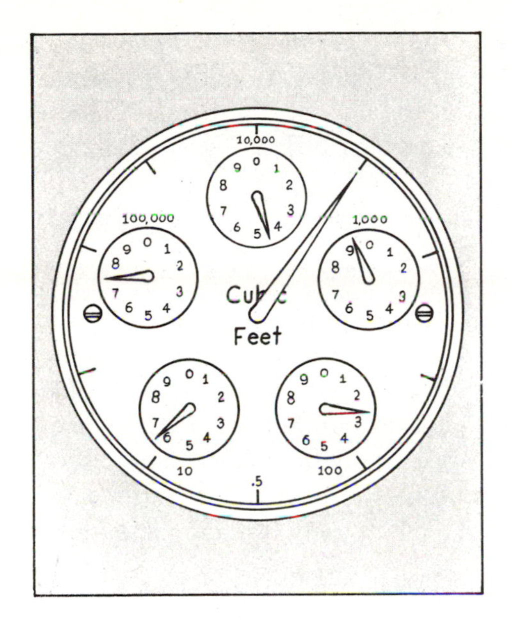

Clockwise dials

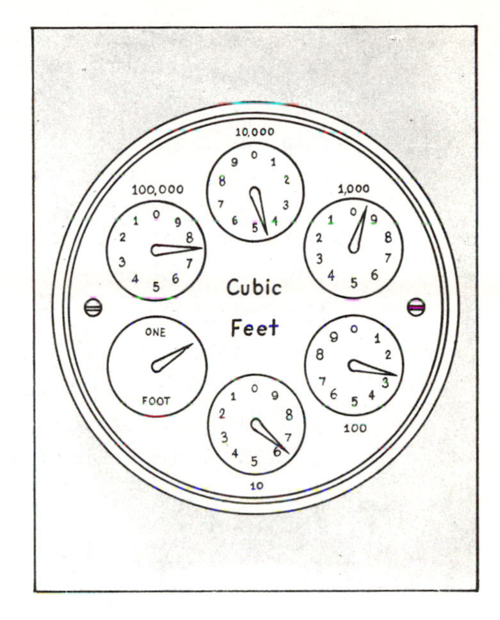

Clockwise and counterclockwise dials

Fig. 1-8 Dial meters.

number of gallons used over a period of time by the washing machine, the lawn sprinkler system, or the filling of the swimming pool, provided the meter is correctly read.

Dial meters are manufactured in two types: the direct reading type which gives the total gallons consumed at a glance, and the cumulative reading meter which has a number of dials that must be read separately, then combined to find the total cubic feet used. A cumulative meter face is shown in Fig. 1-8.

All water meters have a pointer that makes a complete revolution for each cubic foot of water used. To convert cubic feet to gallons multiply the number of cubic feet by 7.5.

One cubic foot of flow is indicated by the long pointer that sweeps the outer edge of the meter face. All of the pointers on the small faces move clockwise. One complete revolution of the large dial represents one cubic foot of water or 7½ gal (28.4 *l*). The number of revolutions on the large dial are then shown on one of the smaller face dials. The number of revolutions on the small dial is recorded on the next small dial face, and so on.

The six-dial meter starts at the dial labelled 100,000; note the smaller of the two digits nearest the pointer in Fig. 1.8. On the

100,000 dial the needle reads 70,000; on the 10,000 dial, the needle reads 4,000; on the 1,000 dial the needle reads 900; on the 100 dial the needle reads 20; and on the 10 dial the needle reads 6. The total reading is therefore 74,926 cu ft or 561,945 gal.

1-5 INSIDE WATER SUPPLY MAIN

The inside main begins at the water meter. There may be more than one supply main. Figure 1-9 shows two main cold water lines—one running along the ceiling in the cellar at one end of the building to feed risers and branches, and another coming off the first main line at right angles to feed other risers and branches. Generally, ¾- or 1-in. (19- or 25.4-mm) pipe is used for the hot and cold water systems and ½-in. (12.7-mm) pipe is used for branch lines to fixtures.

1-6 DOMESTIC HOT WATER SUPPLY

The hot water supply is that part of the water distribution system that brings hot water to the fixtures. The water is heated while passing through a heating coil in the boiler or hot water heater (Fig. 1-10). Both hot and cold water lines should run parallel, spaced at least 6 in. (152 mm) apart to prevent the cold water line from absorbing heat from the hot water line. This requirement can be

Fig. 1-9 The water supply main.

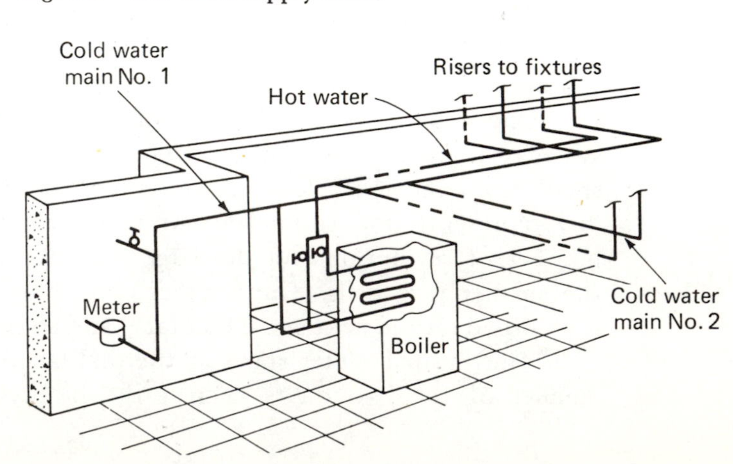

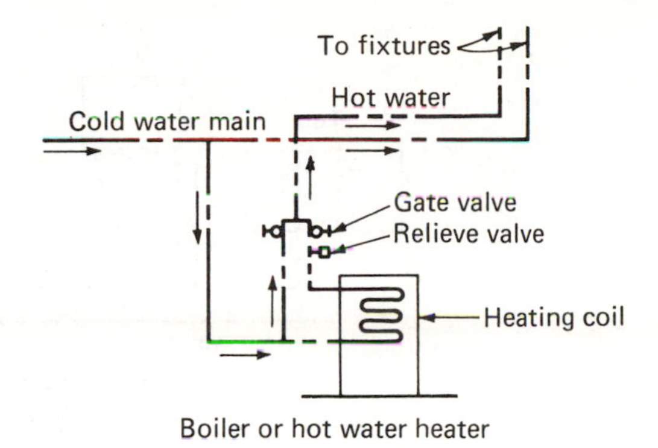

Fig. 1-10 Domestic hot water supply.

ignored if the hot water line is insulated. The hot water line should be run to the left side of each fixture as you face that fixture.

1-7 WORKING LOAD FOR DOMESTIC HOT WATER SUPPLY

The amount of hot water allowed per person ranges from 2 to 10 gal (7.56 to 37.9 *l*) per hour. This amount depends upon the building in which the system is installed. In a school, the hot water allowed per person is 2 to 3 gal (7.56 to 11.4 *l*) per hour. In an apartment building, it is 8 gal (30.3 *l*) per hour per person. In a house, the hot water allowed per person per hour is 10 gal (37.9 *l*). Experience has shown that school buildings average about 25 percent of the rated amount per person. Apartment buildings and houses average about 35 percent per person of the working load.

The working load of the hot water supply for a family of six in a private house is 60 gal (227.1 *l*) per hour. If only 35 percent of the 60 gal are used, then $0.35 \times 60 = 21$ gal (79.5 *l*) are used. Therefore, the heating coil in the boiler must be large enough to produce at least 21 gal of hot water per hour.

1-8 FIXTURE SUPPLY RISERS

Fixture supply risers (Fig. 1-11) are the vertical pipes of the water supply system that feed the branch lines and finally the plumbing fixtures. The tub and lavatory are supplied with hot and cold water, while the water closet (W.C.) is supplied only with cold water.

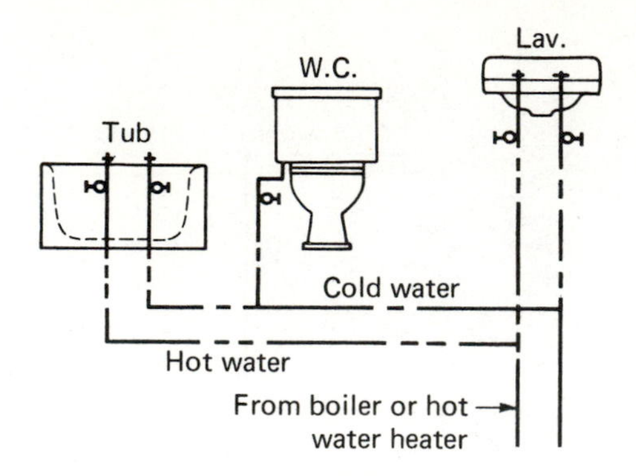

Fig. 1-11 Fixture supply risers.

1-9 USE OF VALVES

The water system is equipped with valves that are placed where they serve the best use. Valves are turned off to repair leaking pipes. There are different kinds of valves used for a water system. One of the most important is the on-off valve, known as the gate valve, shown in Fig. 1-12. It operates by raising and lowering a wedge or disk into the stream of flow.

Others used are the globe valve for throttling the flow of water, while check valves prevent the backup of water. Hot and cold water supply lines to lavatories, sinks, showers, and other fixtures have shutoff valves used when repairs are necessary or when emergencies occur.

1-10 THE GAS HOT WATER HEATER

The installation and maintenance of a hot water heater (Fig. 1-13) is part of the plumber's work, and it is a relatively simple job. However, a hot water heater can also be maintained by a mechanically inclined home owner. Basically the water heater consists of a storage tank and a burner unit at the bottom of the tank, as shown in Fig. 1-14.

Tank capacities typically range from 30 to 60 gal. A 40-gal tank is suitable for a 1250 sq ft (116.12 m^2) house.

The cold water supply comes in at the top of the tank through a dip-tube inside the tank and is routed to the bottom where it mixes with the already heated water. A water temperature dial on the out-

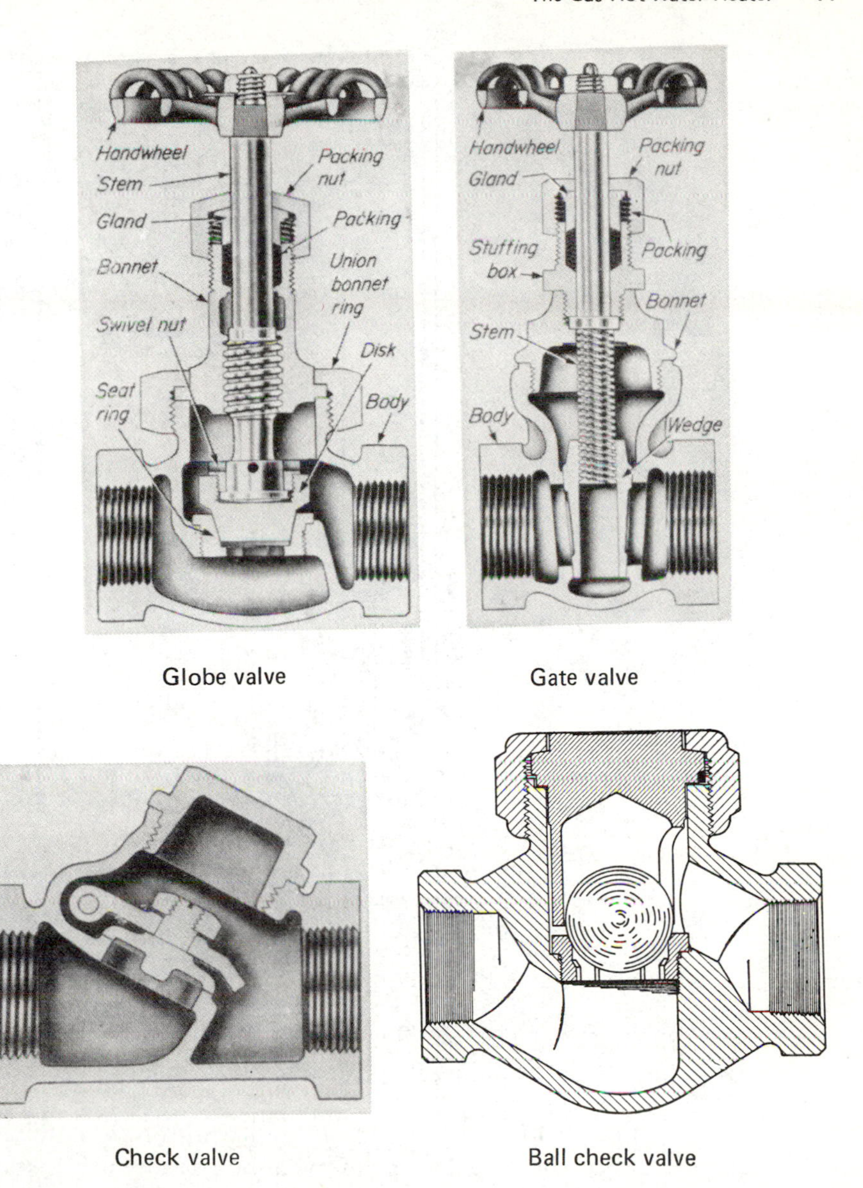

Globe valve

Gate valve

Check valve

Ball check valve

Fig. 1-12 Valves.

side of the tank controls the desired temperature of the water within the tank. The water inside the tank enters the hot water supply line at the top of the tank where it is led under pressure to the fixtures and appliances.

Some of the newer tanks have two magnesium anodes (Fig. 1-15) whose purpose is to attract any particles in the water and

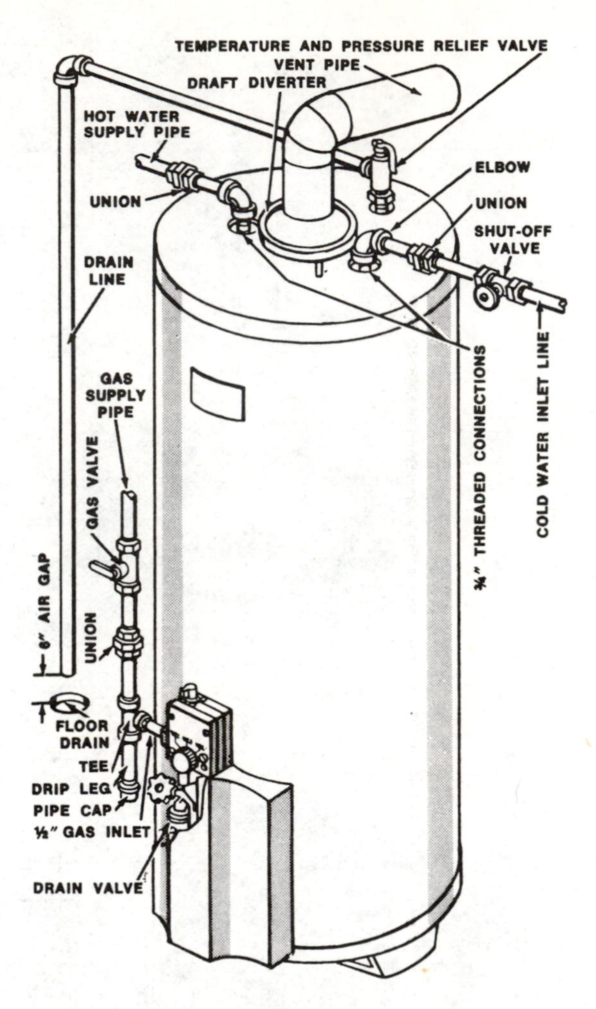

Fig. 1-13 Gas hot water heater.

thus help to keep the tank corrosion-free, adding to the life of the water heater. The primary anode is already installed in the top of the tank when it is first acquired. The secondary anode is not installed but comes with the new heater and is packed inside the water heater carton. The anode is connected to a short piece of pipe which must be installed in the hot water supply outlet of the tank.

When you install the secondary anode, apply paint-joint compound to the threads before placing it in the outlet. Tighten the pipe with a pipe wrench. Most manufacturers will not guarantee the tank if the anodes are not installed.

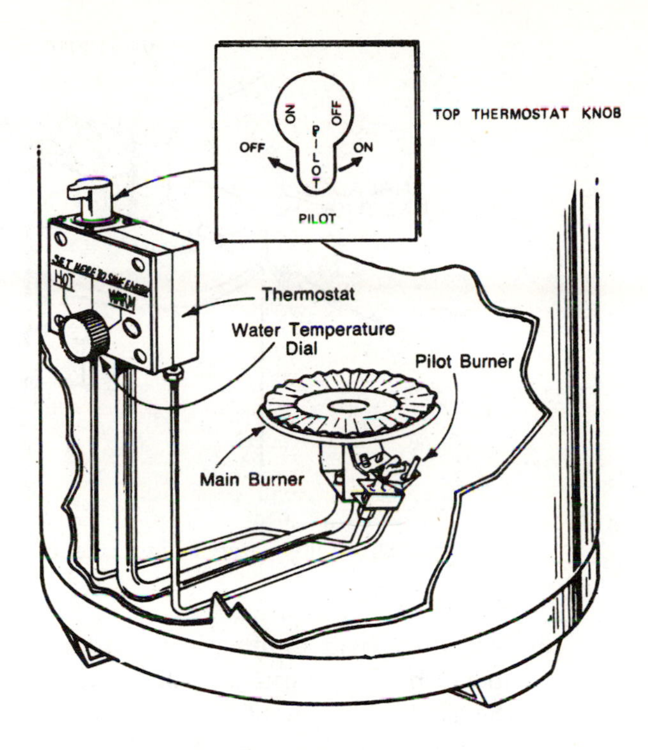

Fig. 1-14 Gas burner unit.

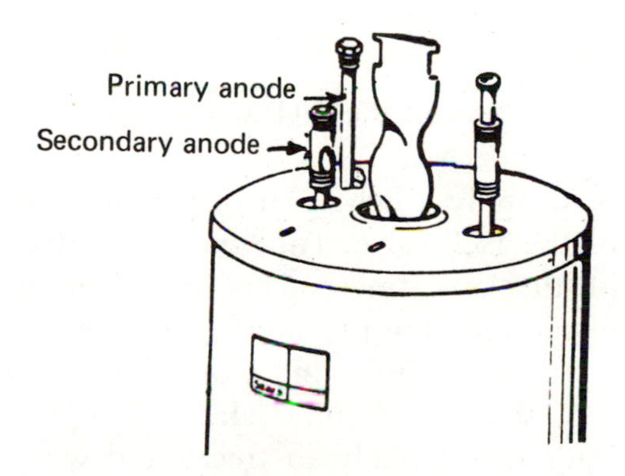

Fig. 1-15 Anodes in tank.

1-11 TEMPERATURE AND PRESSURE (T&P) RELIEF VALVE

The temperature and pressure relief valve must be installed as shown in Fig. 1-16. The sensing element of the T&P valve must extend into the water at the tank side or tank top. The T&P valve is a safety

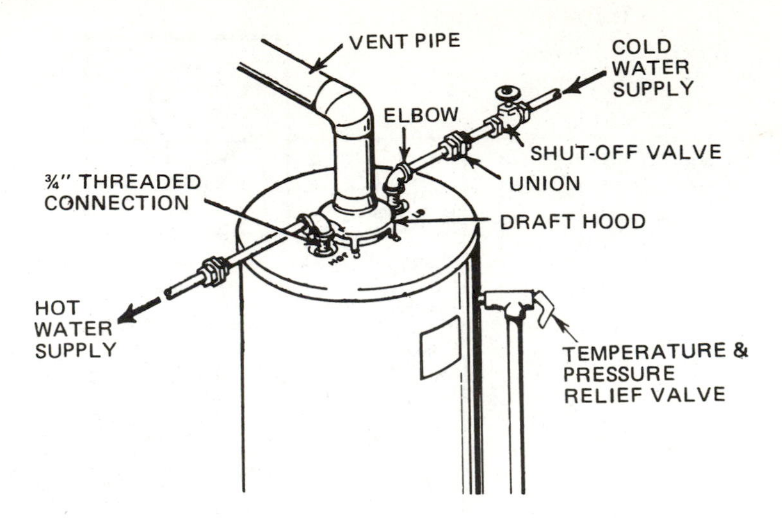

Fig. 1-16 Temperature and pressure relief valve.

device which will automatically drain water if there are excessive temperatures or pressures in the heater. This is the normal function of the T&P valve. Should water damage occur due to an improperly connected drain line, the seller cannot take the responsibility of such damage.

Installation of the T&P Valve

Apply a light coat of pipe-joint compound to the male threads of the T&P valve. Turn the T&P valve into the opening near the top of the tank finger tight and then tighten further with a wrench, being careful not to overtighten the valve.

Run a drain line pipe from the T&P valve outlet to an open tub, sink, or drain. Make certain the place where the drain pipe leads is not likely to freeze and where water will not damage anything. The drain line must be the same size as the T&P valve opening, and its end must be open and unthreaded. Damage could result if the T&P valve opens and the drain line is plugged or frozen. Leave about 6 in. (152 mm) of air gap between the end of the pipe and the draining point.

Note: Do not install a shutoff valve, plug, or cap in the T&P valve drain line.

Installing Copper Connectors

When copper connectors are used as shown in Fig. 1-17, no cutting or threading is needed. The copper connectors replace nipples, elbows, tees and adjust to any alignment. Excessive bending of the connector must be avoided to prevent hardening of the copper. The installation of the copper connectors is relatively easy for the home owner, provided the following steps are carefully observed:

1. Shut off the gas burner or electricity. Turn off the water at the shutoff valve on the incoming water line.
2. Bend the connector into the approximate shape to join the cold water supply to the inlet nipple of the water heater. Bend another connector for the hot water outlet.
3. Make sure all connections are in perfect alignment and

Fig. 1-17 Flexible copper connectors.

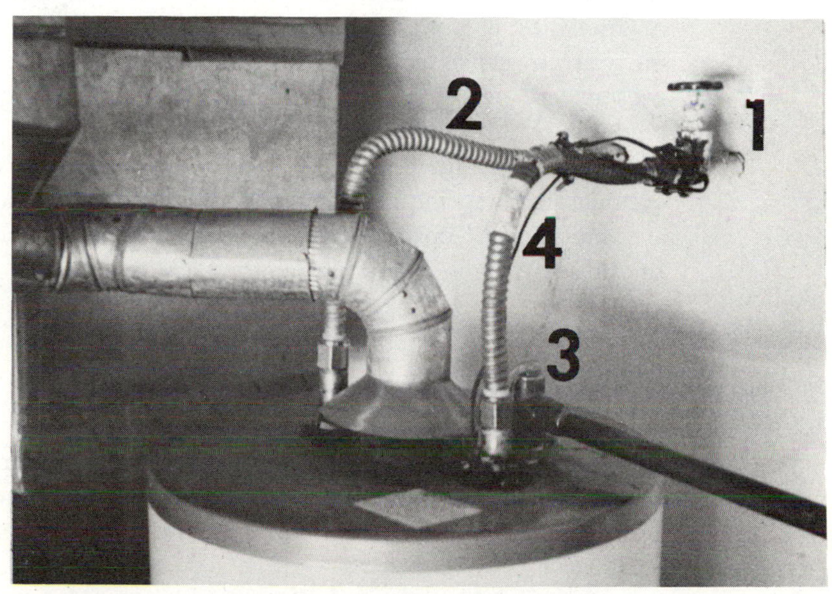

1. Cold water valve
2. Copper connectors
3. Temperature and pressure relief valve
4. No. 10 wire

finger tight before tightening with a wrench. Do not over-tighten. Excessive torque is not necessary.

4. Plumbing water lines are sometimes used to ground electric currents. Since the copper connector is self-insulating, it breaks the electric path. To maintain continuity of ground, the connector must be bonded or jumped. Use approved clamps and connect No. 10 wire from the pipe on one side to the pipe on the other side.
5. Finally after the gas pilot has been lit, turn on the water supply and test the installation.

Electrolytic Action

When different metals or alloys are joined or brought close or into contact with each other and moisture is present, an electrolytic action starts; that is, an electric current begins to flow from one metal to the other and in time one of the metals is eaten away while the other remains intact. How long this takes depends on the amount of moisture present. Even the humidity in the air can be enough moisture to cause electrolytic action. In or near the ocean or in the presence of water containing minerals, electrolytic action can become much more intense.

It is not uncommon for water tanks hooked up with copper connectors to spring leaks less than 3 yr after installation. A greenish accumulation of corroded material, at the end of the copper tube at the brass fitting is a sign of a pending leak.

1-12 PROPER CARE OF THE WATER HEATER

From time to time the home owner must perform two routine chores regarding the efficient operation of his or her water heater. One is clearing the tank of sediment; the other is checking the tank's relief valve.

Clearing the Tank of Sediment

Sediment is the result of rusty or alkaline impurities in the water which come into the hot water tank from the water main. If they accumulate inside the tank, they block the transmission of heat to the water and waste energy. However, sediment settles near the bottom of the tank and can easily be drawn off through the drain valve near the bottom of the tank. How often the tank needs draining depends on the composition of the water and the condition of the tank. Experiment by checking every month until you establish a cycle that allows no more than a pail full of cloudy water to accumulate between drawings.

Checking the Relief Valve

The other maintenance chore is checking the water heater's relief valve. This is an important safety measure. The valve is designed to backup the thermostat. In the event the thermostat should not function properly and too much pressure builds up within the tank, the valve will open automatically and release the overheated water before it can boil into steam and cause an explosion. Check the valve periodically by opening the lever on the valve and allowing some hot water or steam to escape.

1-13 STEPS IN INSTALLING HOT WATER PIPING FOR A HOT WATER HEATER

1. Provide all fittings needed to connect the pipes for both hot and cold water, and drain pipe for the T&P valve.
2. Close the main water valve to shut off all the water going to the house.
3. Open a nearby hot and cold water faucet. This will drain the water pipes. Close the faucets after the pipes drain out.
4. Connect the cold water pipe to the cold water inlet of the heater in the following manner:
   (a) Look at the top of the heater. The water inlet is marked "Cold water supply" and the outlet is marked "Hot water supply."
   (b) Be certain that the dip-tube is in the cold water opening.
   (c) If using copper tubing, solder the tubing to an adapter before attaching the adapter to the water inlet. Do not solder the cold water supply line directly to the cold water connection. It might harm the dip-tube.

(d) The cold water inlet line must have a shutoff valve and union.

(e) To restore cold water service, close the heater shutoff valve and open the main water valve that was shut off earlier.

The hot water heater will work better if the hot water lines are kept short—that is, run the shortest possible way to the fixtures and appliances. Shorter lines will deliver hot water faster with less heat loss. Install the correct valves and fittings needed. Threaded ¾-in. (19-mm) water connections are supplied through the tank top.

Pipe thread compounds, or sealants, as they are often called, are made of a filler material held together by a grease, oil, resinous, or plastic binder. Linseed oil is used as a binder in the preparation of some thread compounds. Calcium carbonate, silicates, lead, or barium oxide powder are suitable for many applications. Barium oxide is particularly conducive to chemical inertness within the compound.

1-14 GAS PIPING

The gas piping for a hot water heater is connected as shown in Fig. 1-18. Black pipe and fittings ½ in. (12.7 mm) are recommended for the gas line to the thermostat. Use a pipe thread compound suitable for the type of gas used. It is important to note the following points when installing the gas piping to the hot water tank:

a. Turn off the gas supply at the meter.
b. Remove all burrs, oil, and old pipe thread compound from the gas pipe. Be sure the inside is clean. Anything left in the gas piping can and will work its way into the thermostat and ruin it.
c. Gas pipes must be tight. Support the thermostat with a wrench when attaching the pipe. This prevents too much pressure on the thermostat when connecting the gas piping.
d. Be sure to check all gas pipes for leaks before lighting the heater. Use a soapy water solution, not a match or open flame. Be sure water is in the tank before lighting burner.

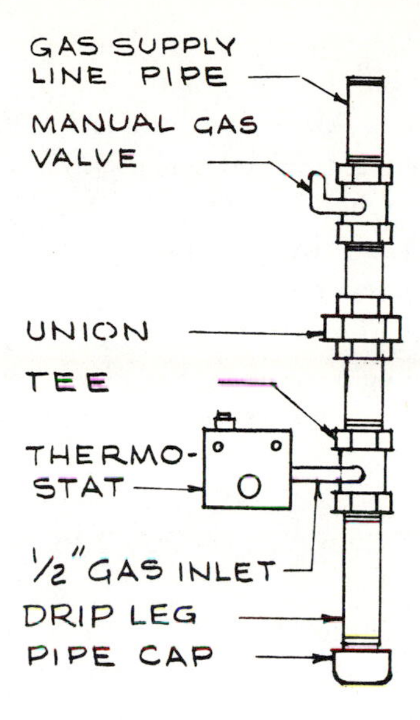

Fig. 1-18 Gas piping to thermostat.

1-15 ELECTRONIC IGNITION SYSTEMS

Instead of using the usual gas jet ignition to light the gas burner, a new automatic system, known as an *"electronic ignition system"* (Fig. 1-19) is often used. When the room thermostat calls for heat,

Fig. 1-19 Electronic ignitor.

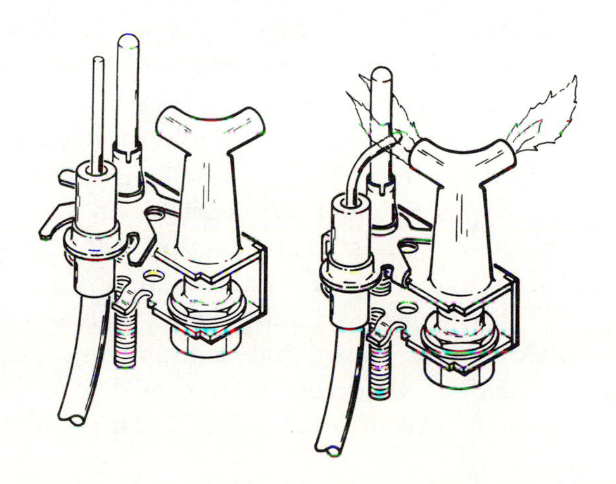

the electronic current flows to an ignitor circuit that opens a pilot valve that brings gas to the pilot. A spark is provided to light the pilot gas. After the pilot is lit, the sparking stops. Flame is provided through a thermal sensing element and gas is furnished to the main burner where it will be ignited by the pilot.

When the thermostat is satisfied and the contacts open, the main burner and the pilot will go off. After a short period of time (about one minute), the system will reset itself in preparation for the next cycle on a demand for heat. In some states the electronic ignitor system has become mandatory.

1-16 FILLING THE TANK WITH WATER

When filling the tank with water, (1) close the heater drain valve, located below the thermostat beneath the access cover; and then (2) open the cold water supply valve to the heater.

Note: This valve must be left open when the heater is in use.

(3) Open a hot water faucet in the kitchen, bath or laundry; and (4) fill the heater tank until water runs out of the open hot water faucet. This will let out the air in the tank and piping. Close the faucet after the water comes out. (5) Check all the new piping for leaks. Fix them as needed and then light the unit.

1-17 DOMESTIC WATER PIPES

Copper pipe is manufactured in two types. One type, known as plumbing tube, includes types K, L, M, and DWV. The other is ACR tube for air conditioning and refrigeration field work.

Each type represents a series of sizes with different wall thicknesses. All types are available in hard tempered straight lengths of 20 ft (6.1 m).

Types K and L are available in soft temper or in an annealed temper in 20-ft straight lengths or in coils, in sizes from ⅜ in. to 6 in. (9.5 to 152 mm) in diameter.

The annealed tubing (types K and L) as well as types M and DWV can be joined by soldering or brazing capillary fittings, or by welding. Soldering with capillary fittings is perhaps the most common way of joining most copper tubing, and is shown in Fig.

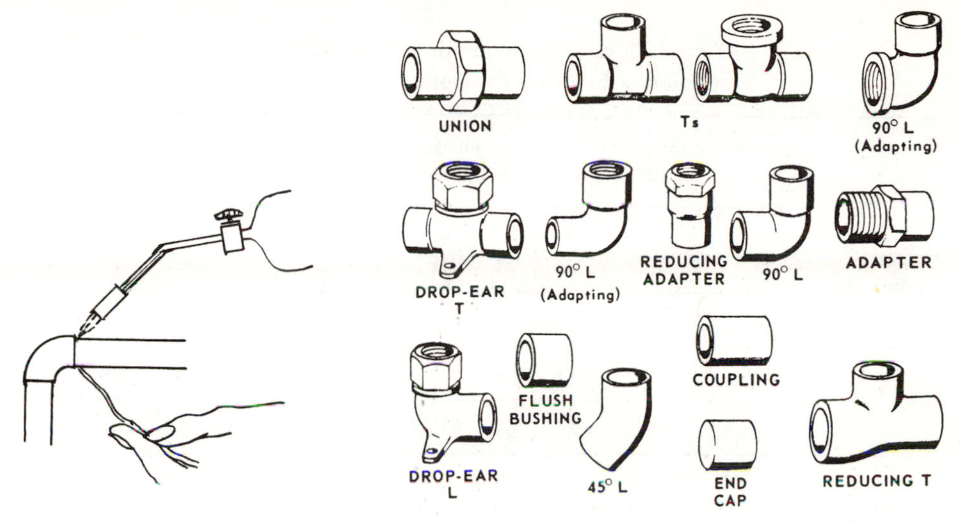

Fig. 1-20 Soldering joint with capillary fitting.

Fig. 1-21 Examples of various solder fittings used with copper pipe.

1-20. Soldered joints are used for water lines. Examples of various solder fittings used with pipes are shown in Fig. 1-21. Available copper tubing sizes are given in Table 1-1, in both inches and metric diameters.

TABLE 1-1

Available Nominal Copper Tubing Sizes

Inch	3/8	1/2	3/4	1	1 1/4	1 1/2	2	2 1/2	3	4	5	6
Metric (mm)	9.52	12.7	19	25.4	31.8	38.1	51	63.7	76	102	127	152

Table 1-2 shows the linear feet of type L copper tubing with soft solder sweat fittings that can be installed per 8-hr, two-worker day. The footage is also given in metres.

Brass pipe with screwed brass fittings is also used for domestic water piping, but of the different types of pipe that can be used, it is the most expensive. Some building codes require that it be used where water is particularly corrosive. Available brass pipe diameters with screwed brass fittings, and galvanized steel or wrought iron

TABLE 1-2

Type L Copper Tubing with Soft Solder Sweat Fittings Principally for Domestic Water Lines

Feet and metres of pipe (tube) lengths and number of fitting units that can be installed per 8-hr, two-worker day

Nominal pipe size													
	in.	3/8	1/2	3/4	1	1 1/4	1 1/2	2	2 1/2	3	4	5	6
	mm	9.52	12.7	19	25.4	31.8	38.1	51	63.7	76	102	127	152
Type L	ft	240	240	230	220	200	170	140	110	80	60	50	40
	m	73.2	73.2	70.1	67	61	51.8	42.7	39.5	24.4	18.3	15.2	12.2
Number of Fittings													
Sweat T		20	20	18	16	14	12	9	7	5	3	2	1
Sweat elbow		30	30	27	24	21	17	13	10	7	5	4	2
Sweat adapter cap		35	35	32	28	24	20	16	12	9	7	5	3
Sweat valve		27	27	24	21	18	15	11	8	6	5	3	2

pipe with galvanized, malleable screwed fittings are shown in Table 1-3.

Galvanized steel pipe or wrought iron pipe with galvanized malleable fitting is economical but not too durable, especially in the smaller sizes (diameters), where interior corrosion can reduce the inside diameter and hence interfere with the flow of water. In the long run the galvanized pipe is less economical than copper tube with sweat joints. Examples of fittings for use with galvanized pipe are shown in Fig. 1-22.

TABLE 1-3

Brass, Galvanized Steel, or Wrought Iron Pipe Diameters

In.	3/8	1/2	3/4	1	1 1/4	1 1/2	2	2 1/2	3	4	5	6
mm	9.52	12.7	19	25.4	31.8	38.1	51	63.7	76	102	127	152

Fig. 1-22 Examples of fittings for use with galvanized pipe.

Saran pipe is an organic thermoplastic material which is used as a replacement for metallic pipe. It is resistant to a wide range of organic and inorganic materials. Saran is available in either pipe or tubing. The pipe is supplied in 10-ft (3-m) lengths only and is made in the diameters shown in Table 1-4.

TABLE 1-4

Saran Pipe Diameters

in.	½	¾	1	1¼	1½	2	2½	3	3½	4
mm	12.7	19	25.4	31.8	38.1	51	63.7	76	88.7	102

TABLE 1-5

Polyethelene Pipe Diameters

in.	½	¾	1	1¼	1½	2	3	4	6
mm	12.7	19	25.4	31.8	38.1	51	63.7	102	152

Polyethelene pipe is mostly used for cold water lines where corrosion resistance and ease of installation is desired. It is well suited for sprinkler or farm water systems and is relatively inexpensive. Table 1-5 gives standard sizes.

1-18 ADVANTAGES OF PLASTIC PIPING

Plastic piping has resistance to practically all acids, salt solutions, and other corrosive liquids and gases. It does not corrode, rust, scale, or pit on the inside or outside of the pipe. It does not rot, and it resists growth of bacteria, algae, and fungi that could cause offensive odor or create serious sanitation problems. Its nontoxicity is a most important factor in food, drug, and chemical processing where fluids must be protected against contamination.

Plastic piping is not subject to galvanic or electrolytic corrosion, a major cause of failure when metal pipe is installed underground. Plastic pipe weighs only one-half to one-sixth as much as metal pipe, which makes it easier to handle, join, and install. It can be fabricated by a variety of methods: solvent welding, fusion welding, threading, and flanging depending on the particular job.

Plastic materials generally are classified into two basic groups—thermoplastics and thermosets. The thermoplastics can be reformed repeatedly by applying heat, whereas once the thermoset has been cured, its shape is fixed and cannot be melted down and reshaped for reuse.

1-19 CPVC (CHLORINATED POLYVINYL CHLORIDE, TYPE IV, GRADE 1 Meeting the Specifications of ASTM D 1785-72)

This material is especially useful in piping for handling high temperature corrosives, having a maximum service temperature of 210° F (99° C). It compares favorably to PVC in chemical resistance. Its recommended uses are for hot and cold water lines and for process piping for hot corrosive liquids.

1-20 POLYPROPYLENE

This is the lightest thermoplastic piping material, but has higher strength and better chemical resistance than polyethelene. Polypropylene is an excellent material for industrial drainage piping where acids, bases, and solvents are involved. It has found wide use in the petroleum industry.

1-21 ABS (ACRYLONITRILE-BUTADIENE-STYRENE, Meeting the Specifications of ASTM D 1527-73)

This material has high impact strength, is tough, and can be used at tempertures up to 180° F. It is also used for carrying drinking water, irrigation, and in drain, waste, and vent piping. It can be joined by threading or solvent welding.

1-22 PIPE SUPPORTS AND HANGERS

All plumbing pipe must be supported to prevent sagging. Cast iron pipe should be hung by hangers spaced about 5 ft (1.52 m) apart as shown in Fig. 1-23. Pipes placed under wooden floor joists must be supported by metal straps spiked to the floor joists.

Fig. 1-23 Pipe hanger.

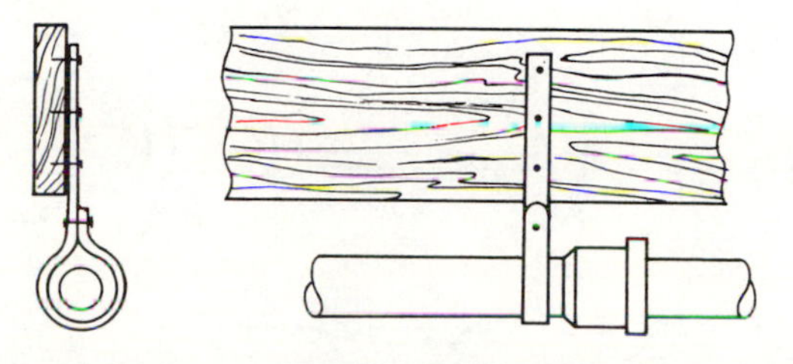

Pipe supported by metal strap

1-23 FIXTURE SHUTOFF VALVES

Almost all plumbing fixtures are now equipped with shutoff valves, which can easily be turned off when a plumbing problem exists. These valves are located under or near the fixtures. The shutoff valve (Fig. 1-24) is attached to the supply pipe, or stub-out as it is called, and to a chrome plated copper connector leading to the fixture. The connector shown is a bayonet type for lavatories.

When attaching the flexible connector to the shutoff valve, be sure to first slip on the coupling nut and the compression ring. Then insert the flexible connector into the valve, seating the com-

Fig. 1-24 Fixture connection.

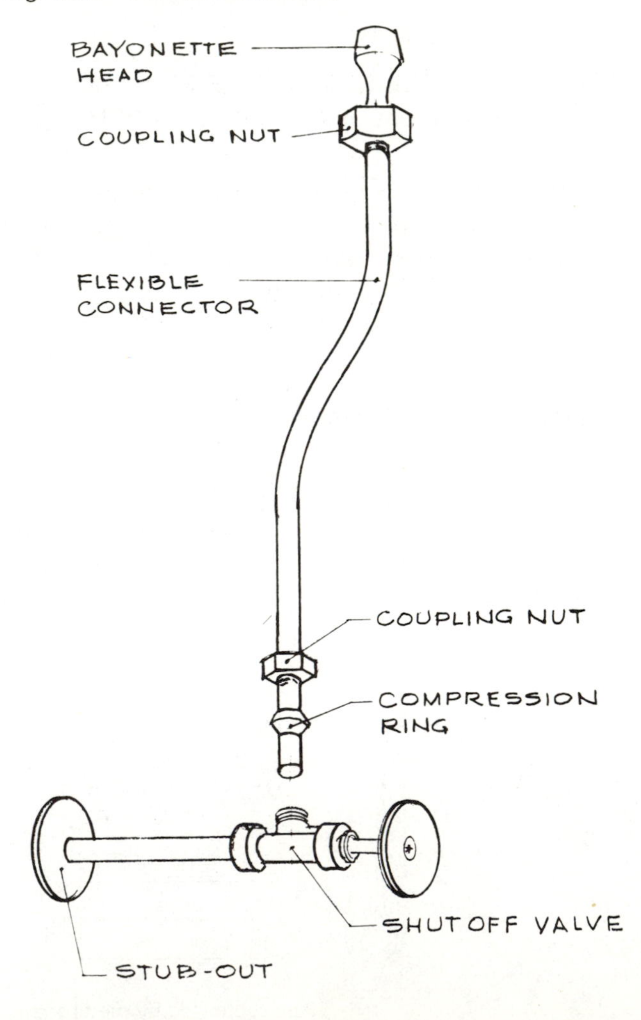

pression ring properly. Then bring down the coupling nut and hand tighten to the shutoff valve. Next, bring the bayonette head into the fixture and bring up and hand tighten the coupling ring. Finish tightening the coupling nut with a basin wrench, and set the compression fitting with an adjustable wrench. Tools are discussed further in Chap. 5.

1-24 HOT AND COLD WATER CONNECTIONS TO LAVATORY

Figure 1-25 shows the proper water connections to a lavatory for both the hot and cold water lines.

Fig. 1-25 Water connections to lavatory.

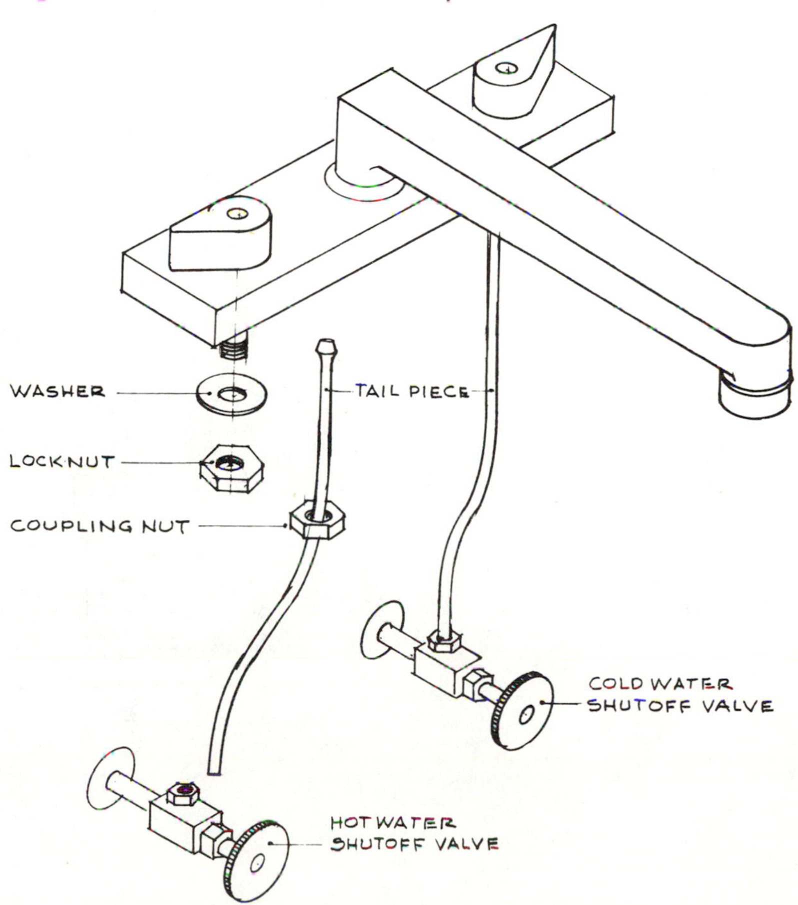

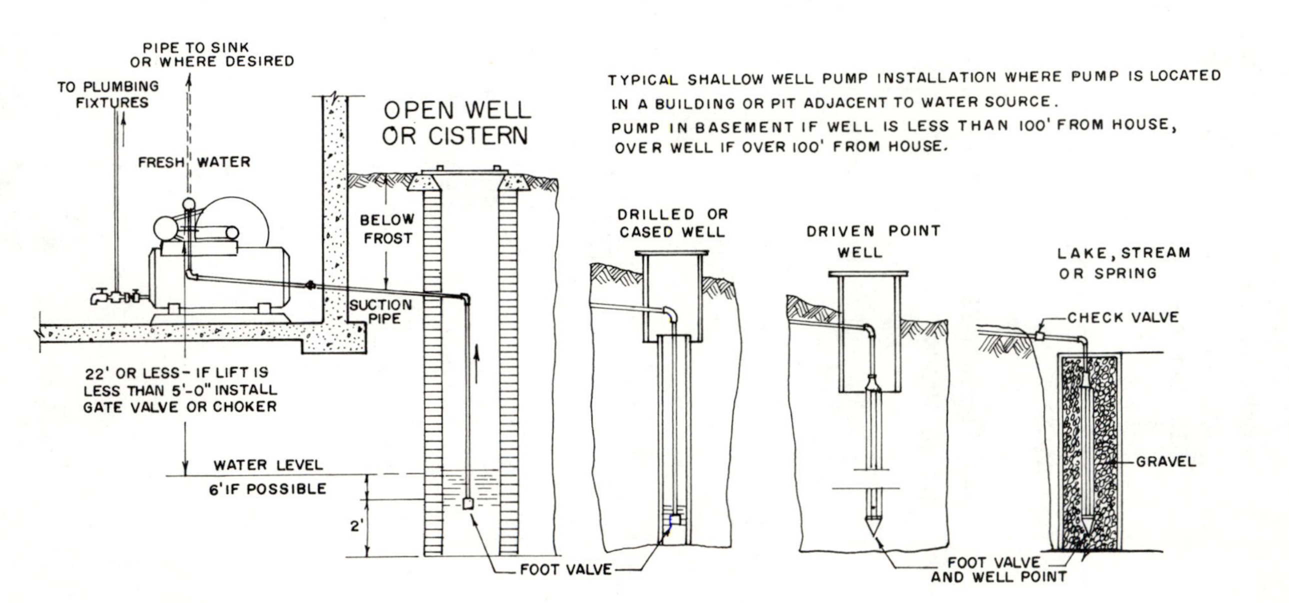

Shallow well pump installation; pump in cellar

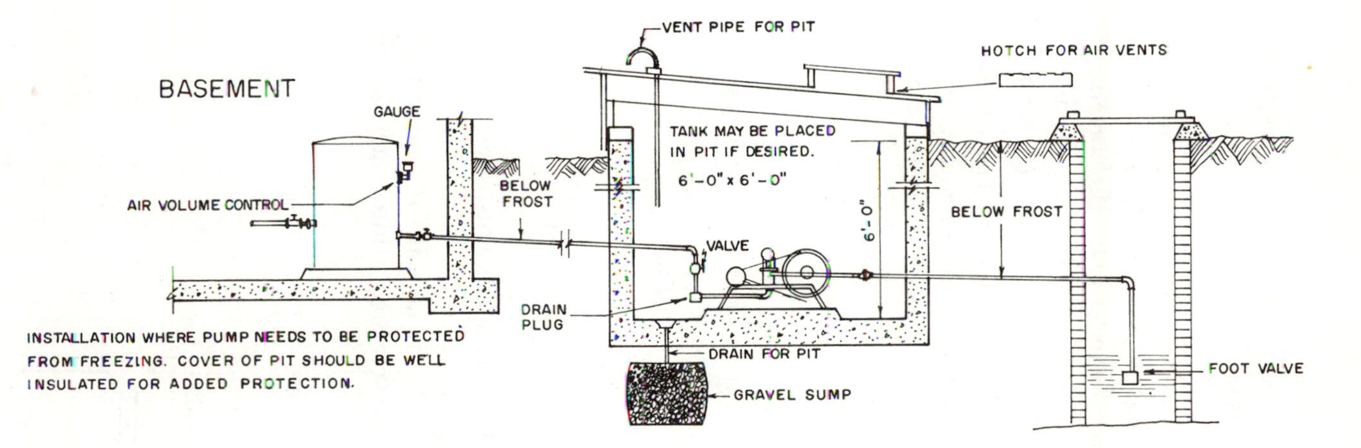

Pump outside of building in insulated pit

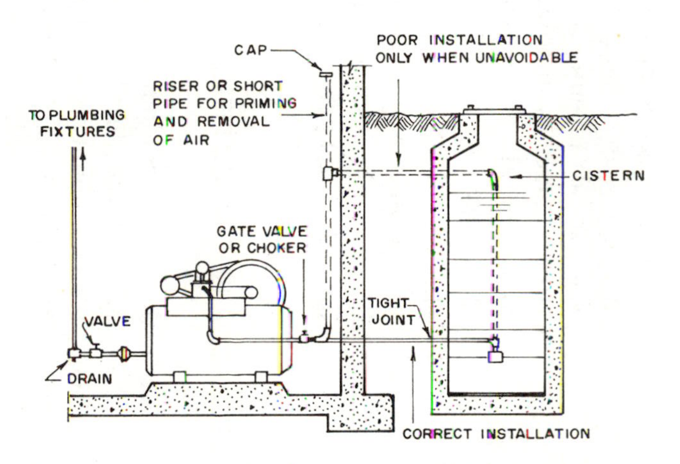

Pump in basement if well is less than 100 ft from house

Fig. 1-26 Well pump data.

1-25 WELL PUMP DATA

For either deep or shallow well systems, three types of pumps are used, as shown in Fig. 1-26. The three types of pumps are:

1. The reciprocating pump operates by means of a piston and cylinder. On the forward stroke a check valve opens, forcing water into the cylinder. On the reverse stroke, the check valve closes, forcing the water collected in the cylinder through the pipe and into the storage tank.
2. The centrifugal pump operates with an impeller. The wheel spins the water, increasing its velocity and pressure. The water spirals up into a casing of decreasing diameter that further increases the water's pressure. As this action is continued, the water is forced through the pipe and into the storage tank.
3. The centrifugal-jet pump sends a stream of water through a small diameter nozzle into a venturi tube, which increases its pressure. From this point, the water is swept up into the centrifugal pump and on to the storage tank.

1-26 MEASURING PIPE FOR CUTTING

A method of determining the length of a pipe is illustrated in Fig. 1-27. Measure the distance between the faces of fittings and add the threaded amount that will go into the fittings. If the face-to-face measurement is 4 ft (1.22 m) and the pipe goes into each fitting

Fig. 1-27 Determining pipe lengths.

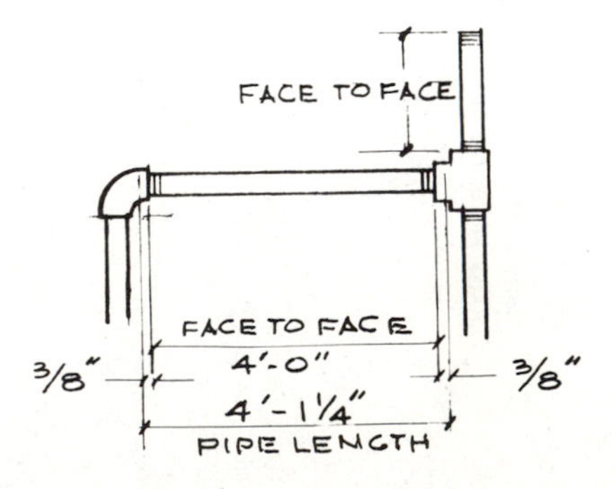

⅝ in. (16 mm), the length of the pipe must be cut to 4 ft (1.22 m) plus ⅝ in. (5.87 mm) at each end or a total pipe length of 4 ft 1¼ in. (1.25 m). The amounts to be added to pipe face-to-face lengths vary.

Galvanized Steel Pipe Diameters

1 -in. pipe (25.4 mm) add ⅝ in. (9.5 mm) at each end or 1¼ in. (31.8 mm)

1½-in. pipe (38 mm) add ¾ in. (9.5 mm) at each end or 1¼ in. (31.8 mm)

2 -in. pipe (51 mm) add ¾ in. (19 mm) at each end or 1½ in. (38 mm)

Cast Iron Pipe

The hubless variety of this pipe is easy to measure, since both ends are plain, fitting into a neoprene washer. If the old hub and spigot type pipe is used, allow for the pipes going into the fittings or hubs. Allow 2½ in. (63.7 mm) for 3-in. (76-mm) pipe and 3 in. (76 mm) for a 4-in. (102-mm) pipe. If the pipe is less than 5 ft (1.5 m) long (cast iron pipe comes in 5- and 10-ft (1.5- and 3.0-m) lengths), use double hub pipe—it has a hub on each end. When it is cut, there is a hub at each end of each cut piece; otherwise, a plain cut-off end would be useless.

Copper

For ringed tubing, use the face-to-face method of measuring, allowing for the distance that the tube will go into the fitting. If flexible tubing with flare fittings is used, use the face-to-face method and allow about 3/16 in. (4.8 mm) for the flare fitting.

1-27 REVIEW QUESTIONS

1. What is the usual house water pressure?
2. What is used to measure the water pressure?

3. Which is the simplest type of well?
4. What is the house water service?
5. What is a corporation stop?
6. One cubic foot of water is how many gallons?
7. Name two types of dial water meters.
8. What is the amount of hot water allowed per person?
9. For a family of six, what is the working load of the hot water supply?
10. Which valve is used to throttle the flow of water?

2 Sanitary Drainage Systems

2-1 INTRODUCTION

There are two types of drainage systems. The system that drains waste water from sinks, tubs, showers, lavatories, and soil matter from toilets is known as the *sanitary drainage system,* and the other system that collects storm water from roofs, yards, and areaways is known as the *storm water drainage system.*

These systems may be used separately, each draining into a public sewer, a private cesspool, or septic tank. When both systems are combined, they are known as a *combination drainage system.*

2-2 SANITARY DRAINAGE SYSTEM

The typical public sewer pipe runs in the center of the street below the frost line and has a minimum fall of ¼ in. per foot. The house sewer taps into the public sewer pipe as shown in Fig. 2-1. The gas and water main lines are usually located as shown.

In order to understand a typical sanitary drainage system it is necessary to know the function of each pipe and fitting within the system. The piping that makes up a sanitary system (shown in Fig. 2-2) is generally as follows:

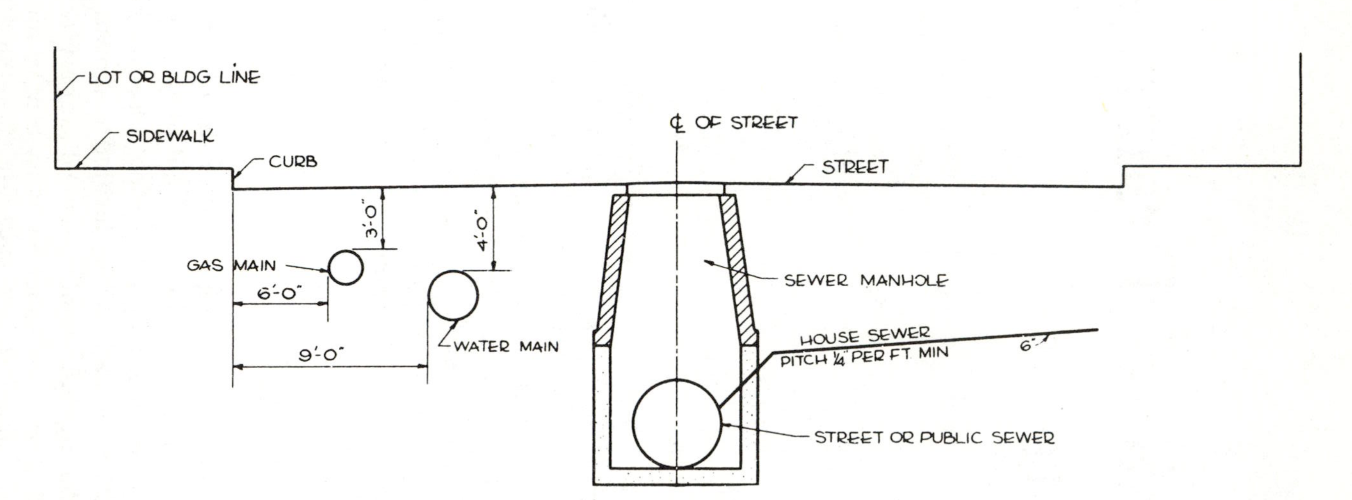

Fig. 2-1 Public sewer and subsurface structures.

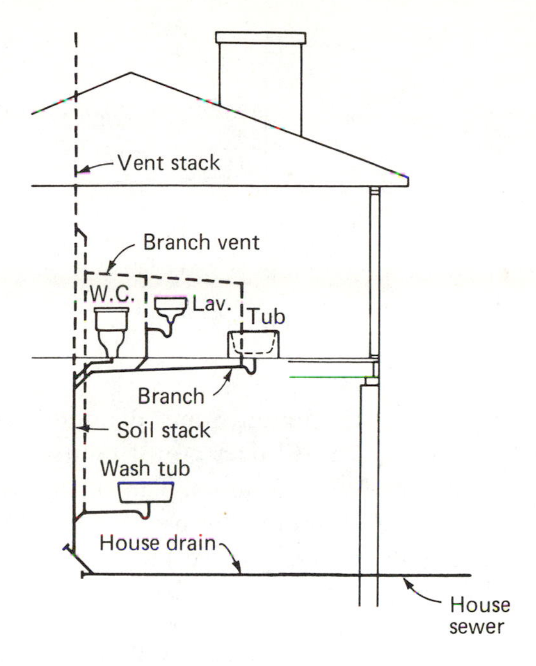

Fig. 2-2 Sanitary system.

1. House sewer
2. House drain
3. Soil, waste, vent stacks
4. Fixture branches
5. Fixture traps
6. Fixtures
7. Fixture vents

2-3 HOUSE SEWER

The house sewer (Fig. 2-3) is that portion of pipe that extends from the public sewer pipe in the street to about 3 to 5 ft (1 to 1.5 m) from the house. The pipe used may be no-hub cast iron, galvanized steel, wrought iron, lead, copper, brass, ABS (acrylonitrile butadiene styrene), PVC (polyvinyl chloride), Transite, or other approved materials depending upon restrictions of local plumbing codes. ABS and PVC installations are limited to residential construction not more than two stories in height.

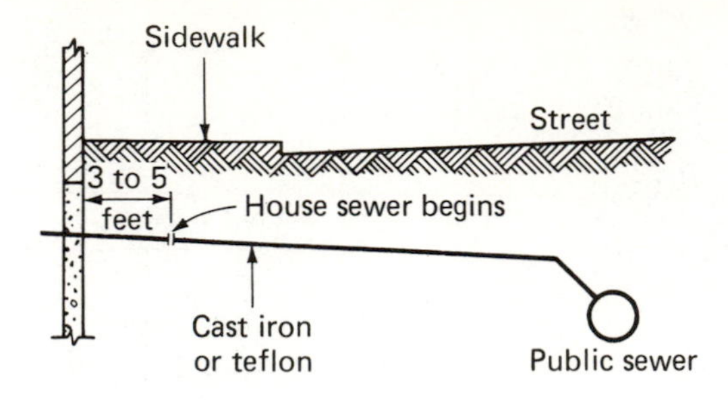

Fig. 2-3 The house sewer.

The minimum pipe diameter for the house sewer is 3 in. (76 mm). Larger sizes are of course required when serving a greater number of fixture units. The strongest pipe that can be used is cast iron, and it should be used wherever strength is required, such as under driveways. Where building codes allow the use of plastic or fiber pipe, they should be used since they are lighter and easier to lift and place.

In place of the old type of house trap and fresh air vent, a back-water pressure valve (BWPV) is installed at the point where the house sewer enters the public sewer. The valve prevents any gases or back-up sewage flow into the house sewer and into fixture traps and fixtures.

If the house sewer line passes within 10 ft (3 m) of a drinking water supply line under pressure, the house sewer line must be leakproof. For cast iron, this means that the joints are to be leaded and caulked, Fig. 2-4, or hubless fittings be used. If plastic or fiber

Fig. 2-4 Joining a bell and spigot.

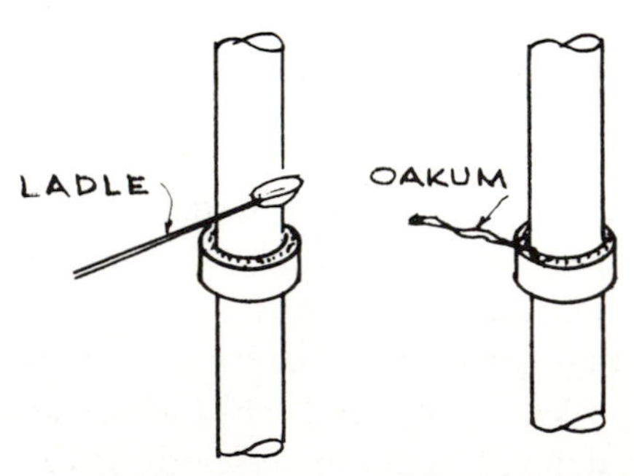

WITH YARNING IRON STUFF OAKUM
INTO JOINT BEFORE POURING LEAD

pipe is used, the joints must be welded and both pipes be solid, not perforated.

2-4 GRADE OF HORIZONTAL HOUSE SEWER DRAINAGE PIPING

Horizontal house sewer piping should have a minimum slope of ¼ in. per foot (20.9 mm/m) toward the point of disposal. Where this is impractical, due to the depth of a street sewer, or the arrangement of the structural features of a building, piping 4 in. or larger in diameter may have a slope of not less than ⅛ in. per foot (10.4 mm/m), if approved by the local administrative authority.

2-5 HOUSE DRAIN

The house drain is the horizontal pipe inside the house into which the vertical soil and waste stacks discharge. It must have a 3-in. (76-mm) minimum diameter and a minimum pitch of ¼ in. per foot (20.9 mm/m), unless similar conditions exist as described for the house sewer.

The location of the house drain depends upon the depth of the public sewer below grade. Since sewage in the house drain flows into the public sewer by gravity, the public sewer must of necessity be deeper below the grade line than the house drain.

The house drain may be under the cellar floor, suspended by metal straps from the floor joists above, or fastened along the cellar wall by metal straps. In a cellarless house the house drain is installed under the concrete slab.

2-6 SOIL, WASTE, AND VENT STACKS

Soil, waste, and vent stacks are the vertical pipes of the drainage system. A distinction must be made between soil, waste, and vent stacks. The soil stack receives the soil matter from the water closet. See Fig. 2-5. The lavatory and tub drain into a waste stack. The soil stack begins at a point where the water closet branch line connects to the stack. The stack above this point becomes a waste

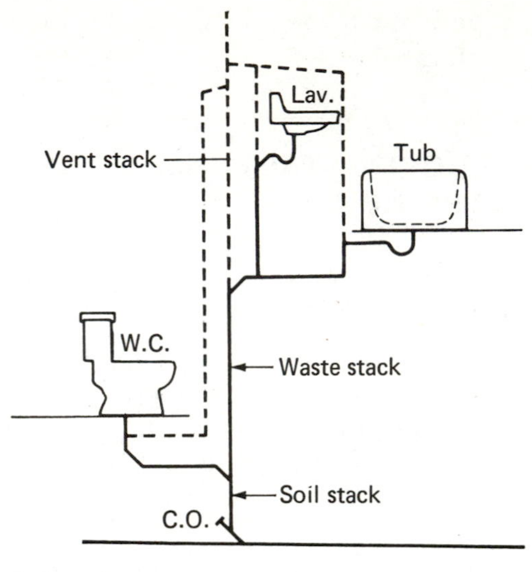

Fig. 2-5 Soil, waste, and vent stacks.

stack. The pipe above the point where the lavatory and the tub drain into the stack becomes a vent stack. Vent stacks are used only to allow air to enter the system.

2-7 FIXTURE BRANCHES AND BRANCH VENTS

The fixture branch is the horizontal pipe running from the fixture trap to the vertical soil or waste stacks. The fixture branch vents are the horizontal piping that run from near the fixture trap, then overhead of the fixture to the vent stack as shown in Fig. 2-6. Branch vents should be graded so that any moisture or condensate that may collect in them can flow back to the branches.

Branch vents are arranged so that waste matter flowing through the fixture branch cannot clog and foul the vents. For this reason branch vents are never connected to the crown of traps. Branch vents should never be taken from fixture branches below the hydraulic grade. *Hydraulic grade* is a line from the high water level of a fixture such as the lavatory (Fig. 2-7) to the branch connection at the soil stack.

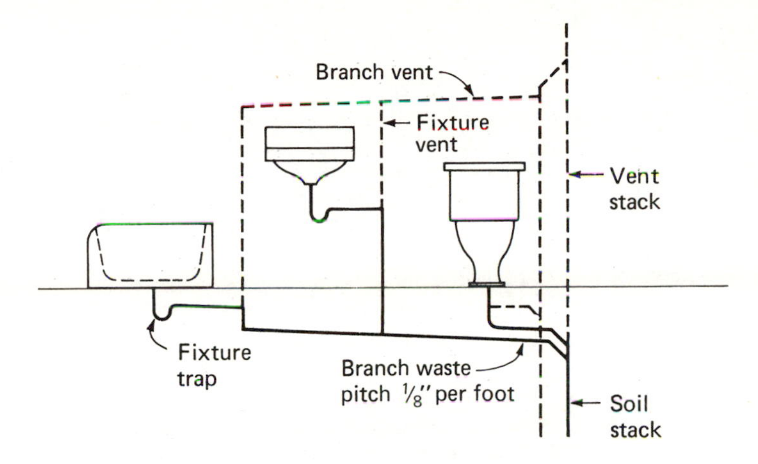

Fig. 2-6 Fixture branches and branch vents.

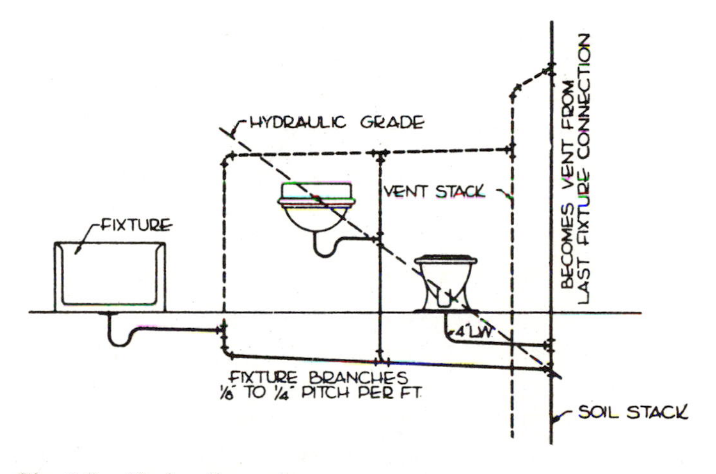

Fig. 2-7 Hydraulic grade.

2-8 FIXTURE TRAPS

Every fixture has a trap. The trap has a water seal, which prevents odors and gases within the pipe from entering the living areas. The trap is directly under the fixture and is U-shaped as shown in Fig. 2-8. The water seal in the trap must not be less than 2 in. (50.8 mm) and not more than 4 in. (101.6 mm) except where a deeper seal is found necessary. Traps must be protected from freezing.

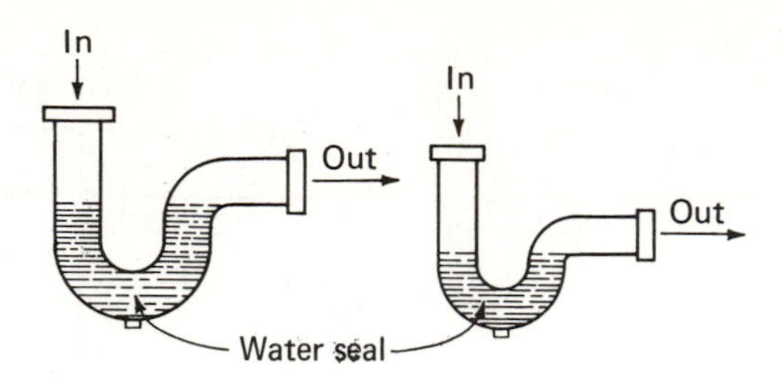

Fig. 2-8 Fixture traps.

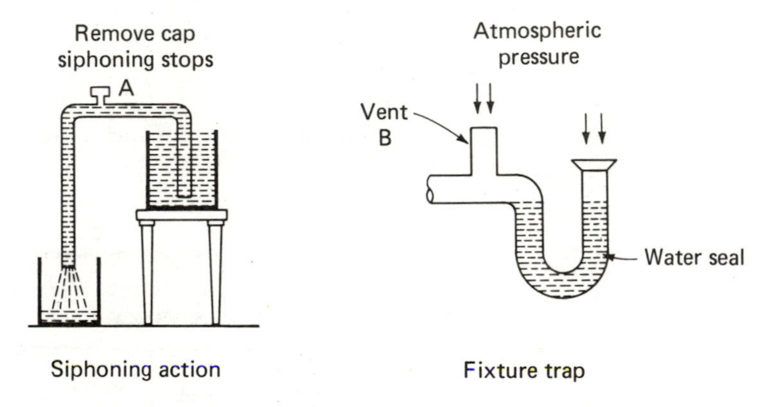

Fig. 2-9 Siphoning action.

Each fixture trap must be vented to prevent *siphonage* of the fixture trap. Siphonage can best be explained by referring to Fig. 2-9. A tube is used for moving liquid out over the top edge of the container. This occurs because of the force of atmospheric pressure acting upon the surface of the liquid. One end of the tube is placed in the liquid and the other end in a container placed at a lower level. The tube must be filled by suction before flow will start. If valve A is opened, the siphoning action will stop. Similarly, a vent at B will stop the siphoning of the trap as soon as atmospheric pressures are equalized.

2-9 FIXTURES

What are fixtures? Plumbers say that kitchen sink as well as a bathroom wash basin, a toilet bowl, a shower, or a tub are fixtures. Figure 2-10 shows a number of fixtures. The common names of

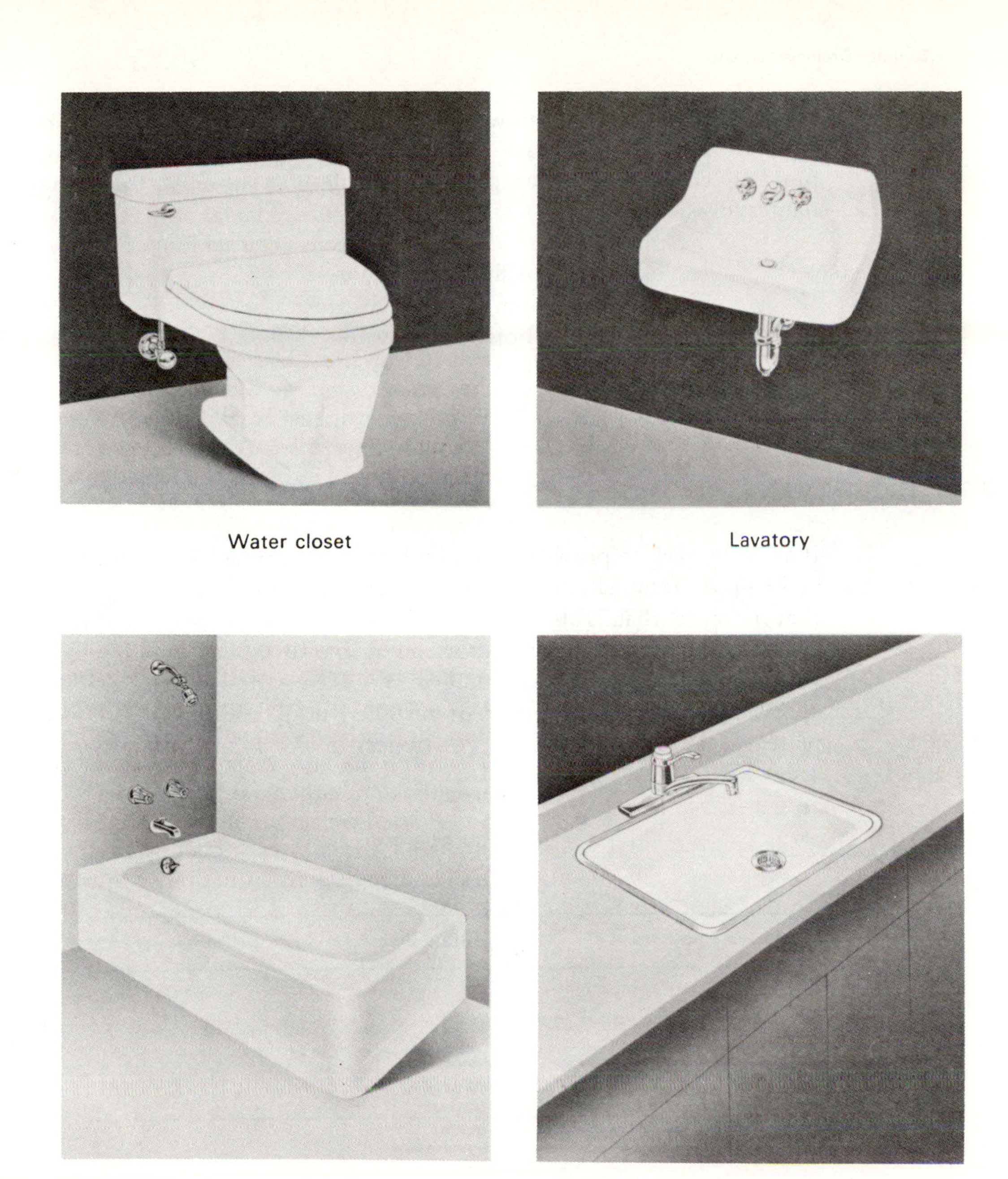

Fig. 2-10 Plumbing fixtures. (Courtesy of American Standard Plumbing and Heating.)

these fixtures are well known, whereas the trade names are not so well known. For purposes of exact meaning and interpretation, building codes as well as plumbing tradesmen use the following names for the fixtures:

Kitchen sink	sink
Bathroom sink	lavatory
Toilet bowl	water closet
Shower	shower stall
Bathtub	tub

2-10 CLEANOUTS

It is customary to provide horizontal drainage piping with a cleanout at its upper terminal, and additional cleanouts on each run of piping that is more than 100 ft (30.5 m) in total developed length. A cleanout is also required in a horizontal line of piping when there is a change of direction exceeding 135°. Cleanouts must be readily accessible, either above grade or installed under an approved cover plate. Cleanout fittings must not be less in size than those given in Table 2-1.

TABLE 2-1

Cleanout Sizes in Inches

Size of Pipe	Size of Cleanout	Threads per Inch
1½	1½	11½
2	1½	11½
2½	2½	8
3	2½	8
4 and larger	3½	8

2-11 TYPICAL SEWAGE DISPOSAL SYSTEM WITHIN THE HOUSE

The plumbing system shown in Fig. 2-11 is for a typical two-story house with cellar. The waste and soil stacks discharge into the house drain that must be a minimum of 3 in. (76 mm) but is, in most

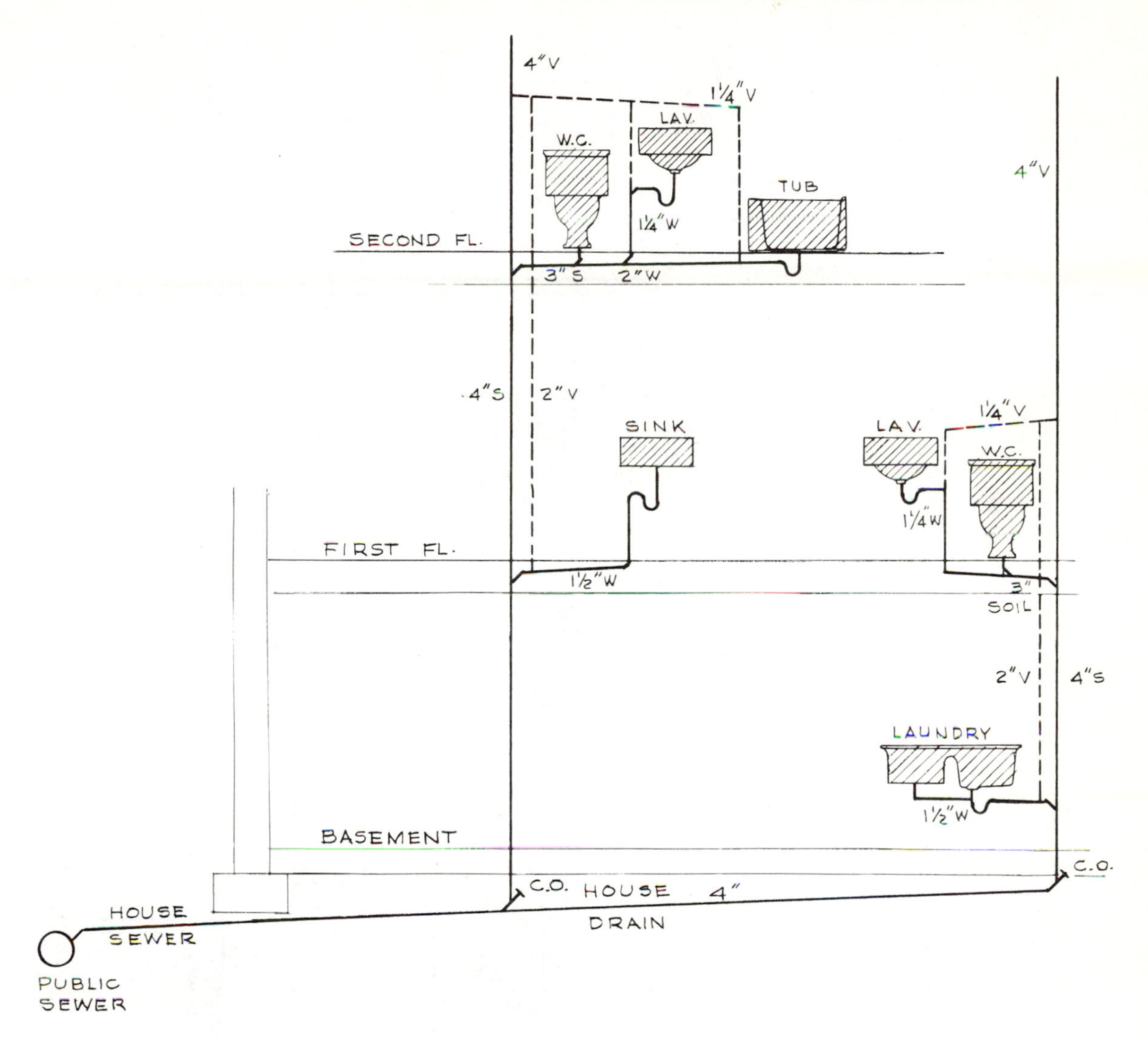

Fig. 2-11 Sewer disposal system.

cases, a 4-in. diameter extra heavy cast iron pipe with lead joints. The house drain is installed under the cellar concrete floor slab.

The house drain connects to the house sewer which slopes ¼ in. per foot to the public sewer.

The house sewer may be of extra heavy cast iron with leaded joints, or, where building codes permit, PVC pipe with rubber gaskets, or no-hub cast iron pipe with cast iron clamps, or pipe made of Transite.

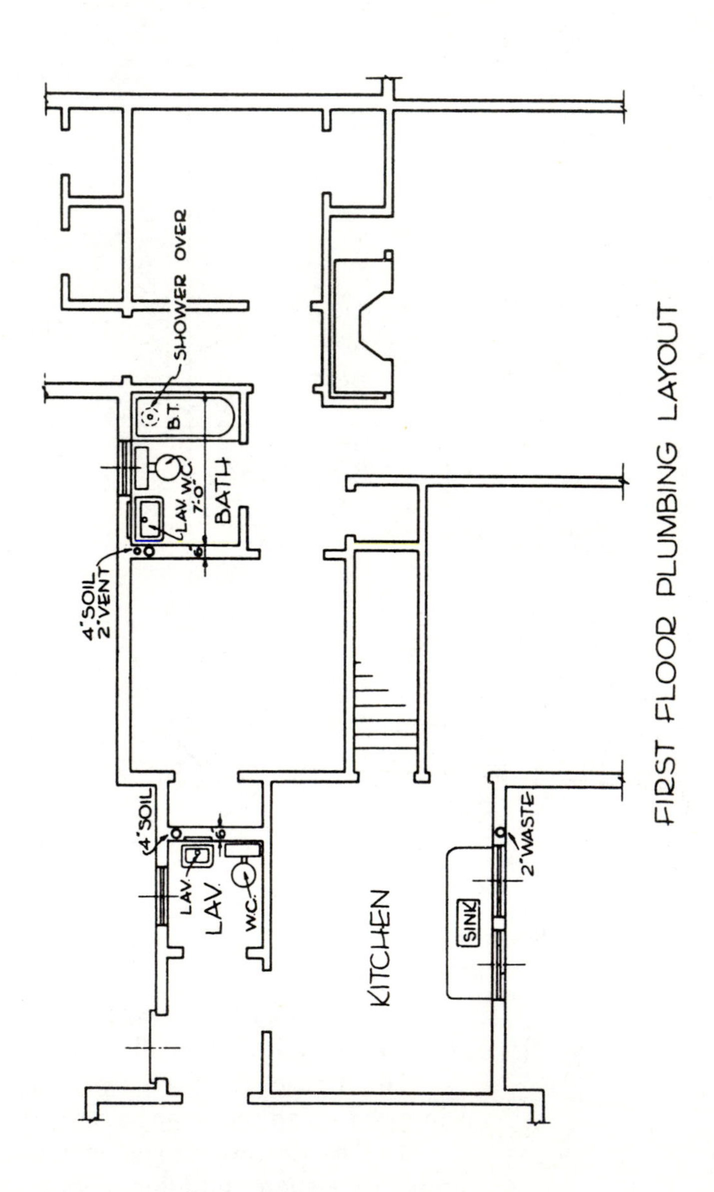

FIRST FLOOR PLUMBING LAYOUT
SHOWER OVER
B.T.
W.C.
LAV.
7'-0"
BATH
4" SOIL
2" VENT
6"
4" SOIL
LAV.
LAV
W.C.
KITCHEN
SINK
2" WASTE

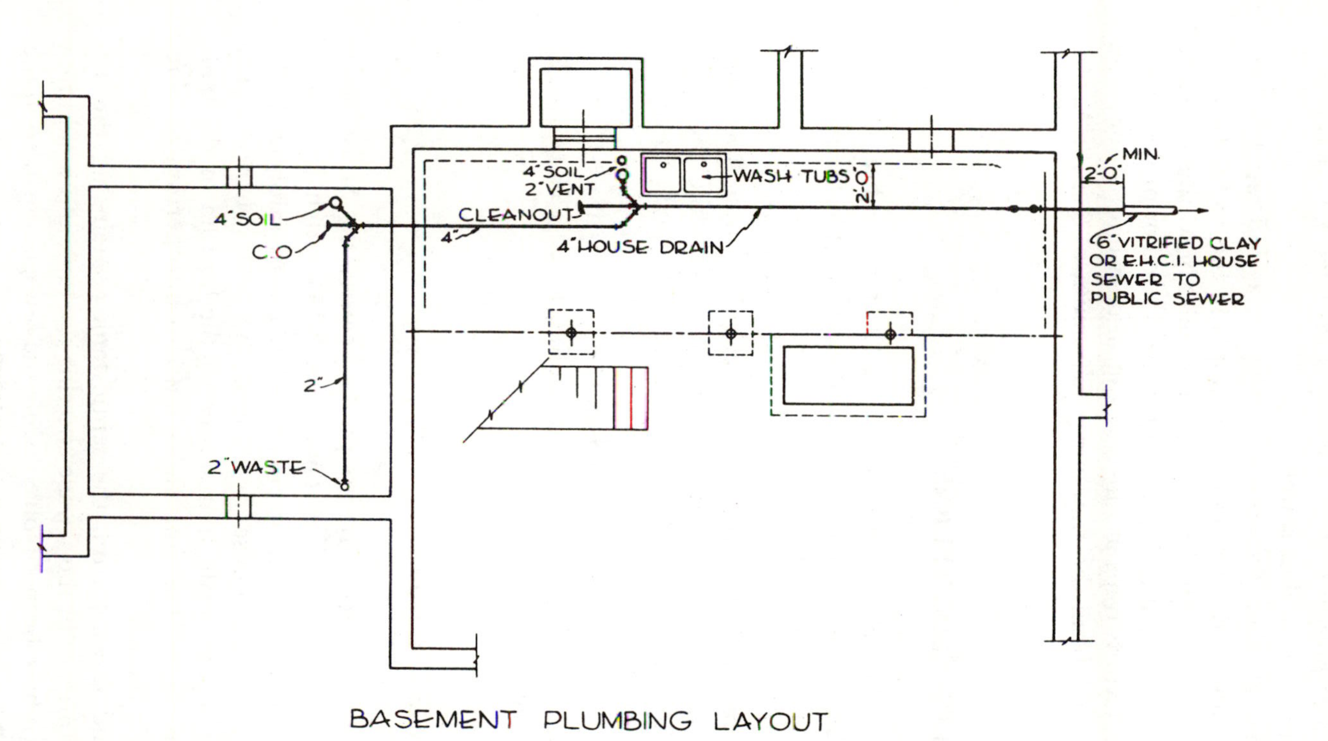

Fig. 2-12 Plumbing plans.

A backwater pressure valve (BWPV) is installed just before the house sewer enters the public sewer.

Note, that all nine fixtures are trapped, as indicated by the U-shaped line under the fixtures. The trap for the toilet (w.c.) is not shown since it is built directly into the toilet bowl. Each fixture trap must also be vented as indicated by the dotted or broken lines. Vents for the fixture traps are installed in branch lines a minimum of 6 in. (152 mm) from the trap and a maximum of 6 ft (2m) from the trap.

2-12 PLUMBING LAYOUT ON FLOOR PLANS

The basement plumbing layout in Fig. 2-12 shows a house sewer which may be of 6-in. (152-mm) vitrified clay, plastic pipe, or extra heavy cast iron pipe leading to a public sewer. The house drain, inside the building, receives soil and waste from the basement wash tubs and the fixtures indicated on the first floor plan. The tub and lavatory drain into the waste stack while the water closet drains into the soil stack. Each fixture is trapped and vented. The kitchen sink drains into a waste stack and is trapped and vented, and the lavatory fixtures are trapped and vented. Figure 2-13 shows a diagramatic riser diagram of the plumbing layout. Note the locations of the cleanouts.

2-13 SEPTIC TANK SEWAGE DISPOSAL

In localities where there are no public sewers, the house sewer drains into a household septic tank (Fig. 2-14) which is made of concrete or salt-glazed tile, or concrete lined with salt-glazed tile. The sewage from the house flows into the tank where bacteria dissolve some of the waste that settles to the bottom of the tank. The liquid on top of the tank flows out and into perforated pipes and disperses in the soil.

The liquid flowing from the septic tank contains all the dissolved matter present in the sewage entering the tank and in addition all solids which have been dissolved while the sewage remained in the tank in contact with the decomposing sludge.

Since most of these dissolved solids are organic waste matter,

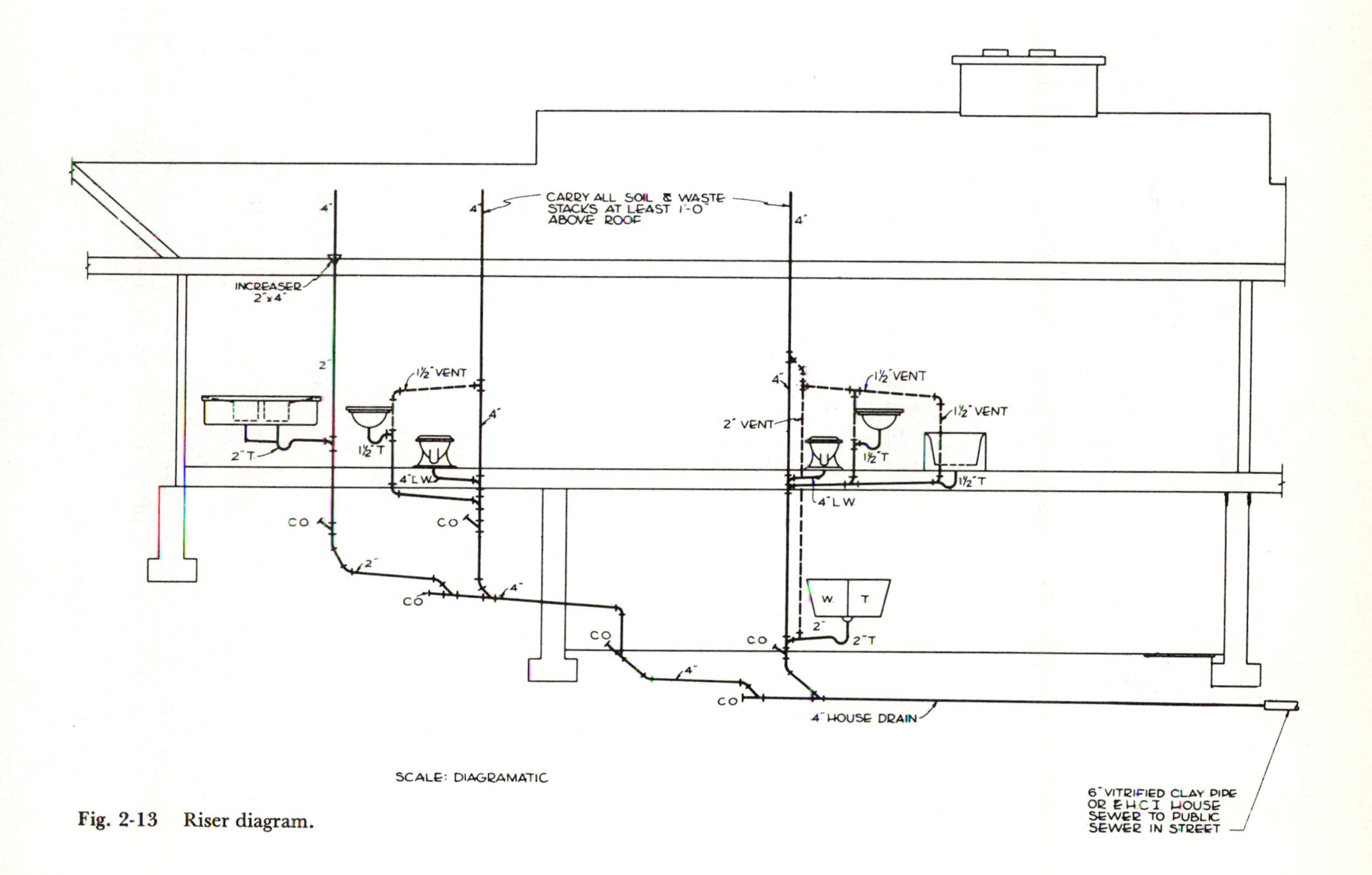

Fig. 2-13 Riser diagram.

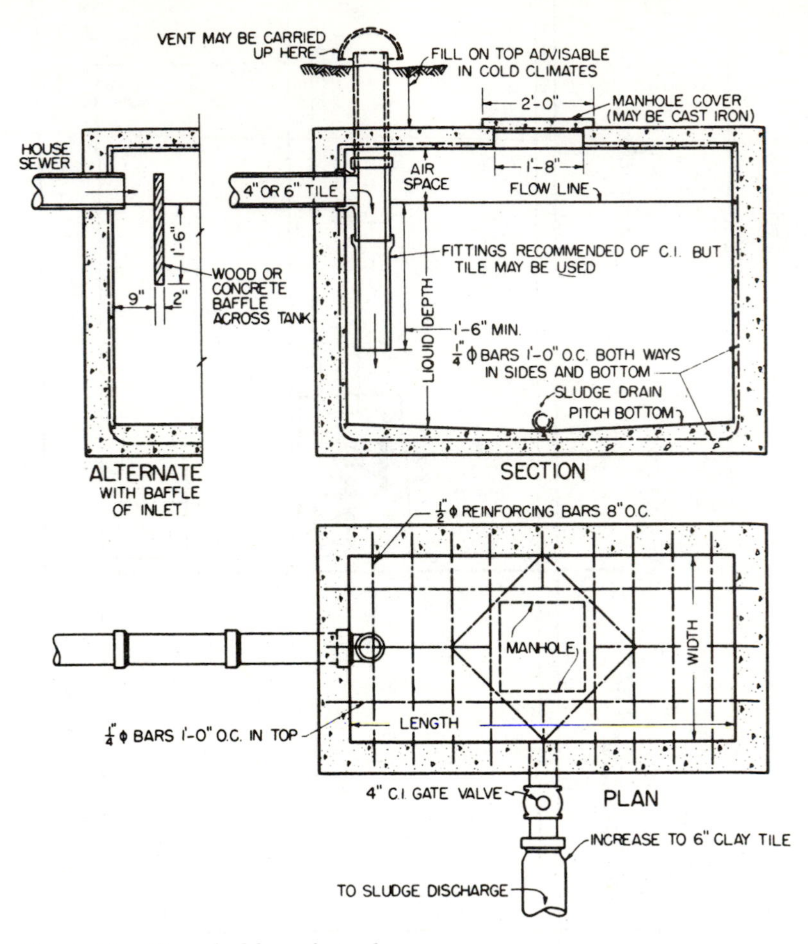

Fig. 2-14 Household septic tank.

they are highly decomposable and become very offensive unless proper precautions are taken for the disposal of the effluent liquid. Contrary to popular belief, germs and bacteria are not completely destroyed in a septic tank, but are carried off in the effluent to the disposal field. Therefore, the effluent should never be discharged on the surface of the ground, into a gutter, or into an open trench. A site must be selected where such disposal will not endanger any water, well, or other source of drinking water. The effluent from the tank is then disposed of by dispersion in the upper layers of soil by means of a system of tile sewer pipes with open joints, in an area called a disposal or irrigation field (Fig. 2-15). Drainage tile is available in 2-ft lengths with a 4-in. inside diameter.

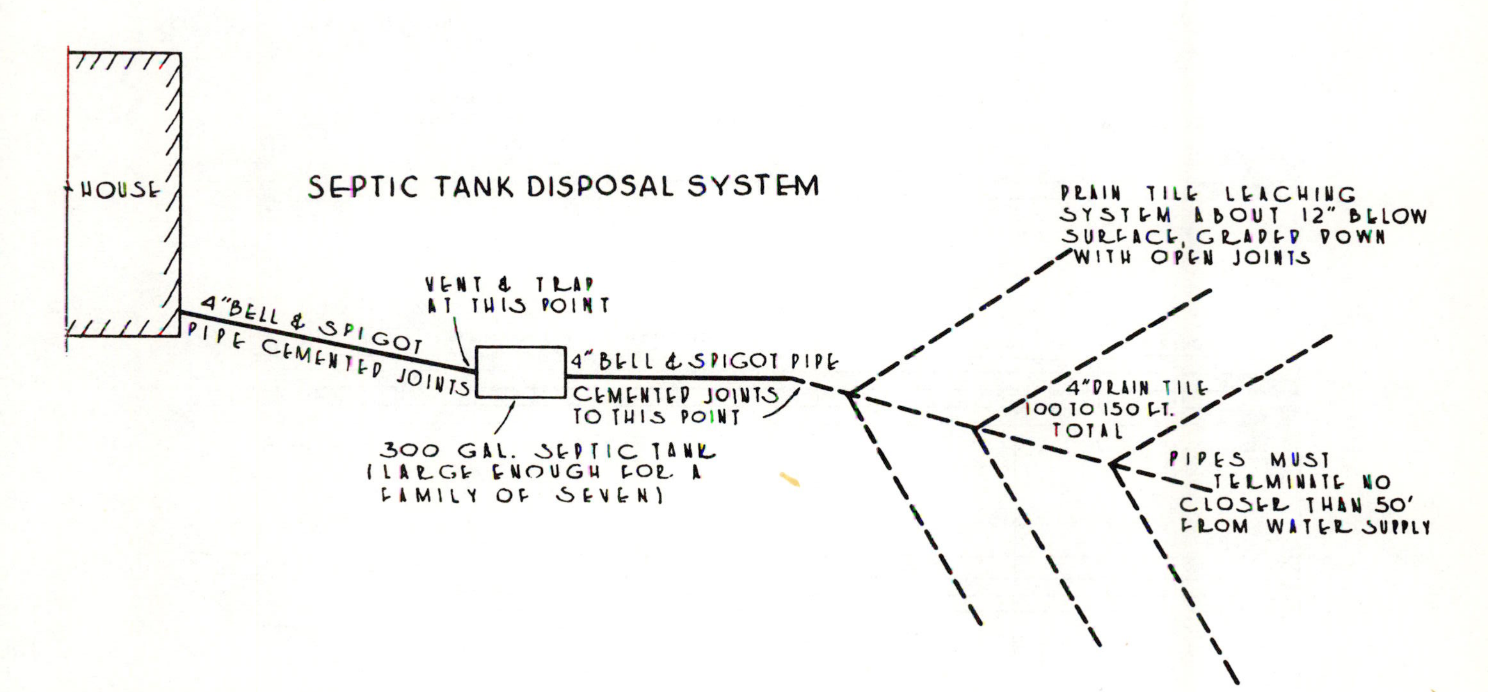

Fig. 2-15 Septic tank disposal system.

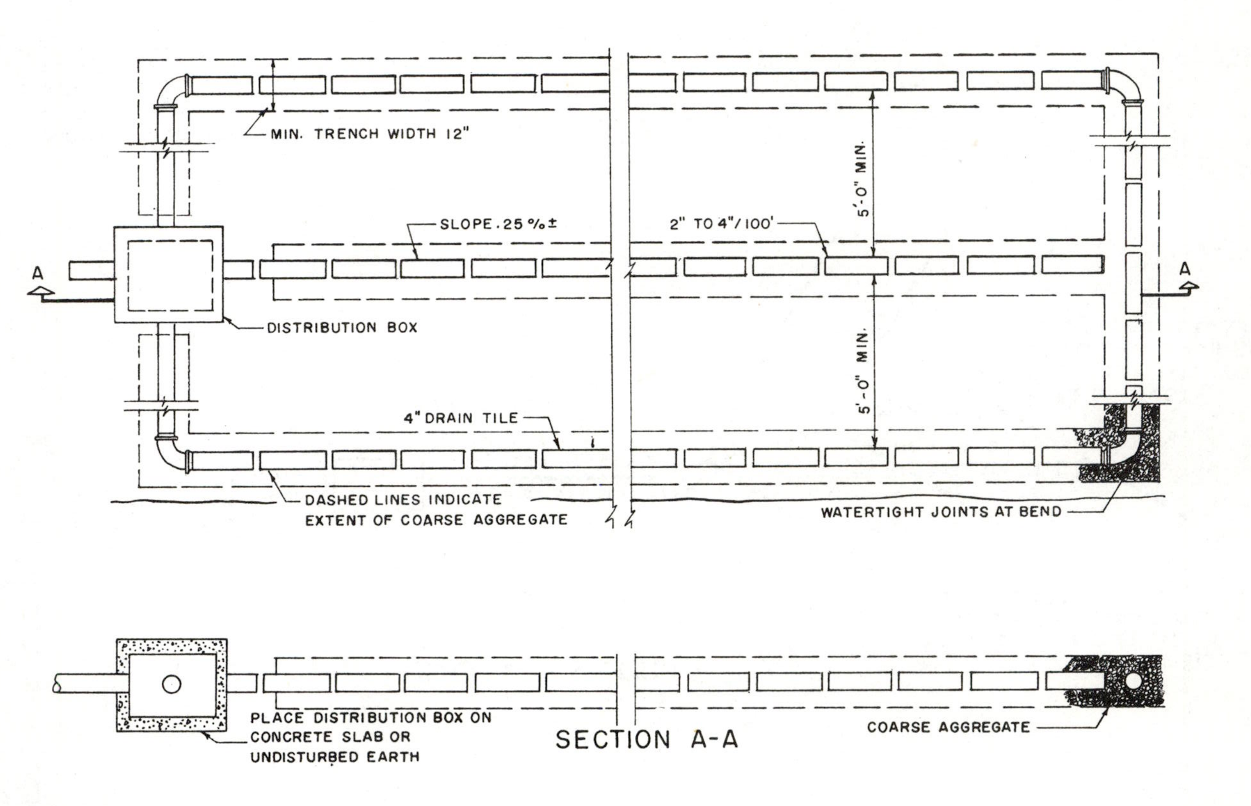

Fig. 2-16 Irrigation field.

The septic tank should not be placed closer than 25 ft (7.6 m) to a building. Pipes in the disposal field should have a grade of 4 to 6 in. (100 to 152 mm) per 100 ft (30.5 m) in porous soils and 2 to 3 in. (51 to 76 mm) per 100 ft (30.5 m) in compact soils. The area of the disposal field should not be less than 600 sq ft ($55.7 m^2$) for up to five persons. For each additional person allow an additional 40 linear feet (12 m) of pipe. The distance between lines in the disposal field should be a minimum of 5 ft (1.5 m). A disposal or irrigation field on level ground is shown in Fig. 2-16. Table 2-2 gives the dimensions of septic tanks for various numbers of persons.

TABLE 2-2

Sewage Disposal System Data

	Capacity of Septic Tank					Concrete Thickness		
Number of persons served	Working capacity (gal)	Length	Width	Air Space	Liquid Depth	Walls (in.)	Top (in.)	Bottom (in.)
1- 4	325	5'-0"	2'-6"	1'-0"	3'-6"	6	4	6
5- 9	450	6'-0"	2'-6"	1'-0"	4'-0"	6	4	6
10-14	720	7'-0"	3'-6"	1'-0"	4'-0"	6	4	6
15-20	1000	8'-0"	4'-0"	1'-0"	4'-0"	6	4	6
21-25	1250	9'-0"	4'-6"	1'-0"	4'-3"	7	5	6
26-30	1480	9'-6"	4'-8"	1'-3"	4'-6"	8	5	6
31-35	1720	10'-0"	5'-0"	1'-3"	4'-8"	8	5	6
36-40	1960	10'-6"	5'-3"	1'-3"	4'-9"	9	5	6
41-45	2175	11'-0"	5'-6"	1'-3"	4'-10"	9	5	6
46-50	2400	11'-6"	5'-9"	1'-3"	5'-0"	9	5	6

Refer to Fig. 2-14.

2-14 FIXTURE UNITS

Fixture units represent the gallons of water per mintue that can be drained from a fixture. For example, a lavatory can drain 7½ gal per minute, or one fixture unit. Table 2-3 is based upon the rate of discharge from a lavatory as the unit is employed to determine fixture equivalents.

TABLE 2-3

Fixture Units

Fixture	Fixture Unit
Lavatory or wash basin	1
Bathtub	2
Laundry tray	2
Sink (except slop sink)	2
Combination fixture	3
Urinal	3
Shower bath	2
Floor drain	2
Slop sink	3
Water closet	6
Slop sink with flushing rim	6
Drinking fountain	½
Dental cuspidor	½
Bathroom unit containing one water closet, one lavatory, one bathtub, with or without shower, or one shower stall	6

Fig. 2-17 Sizing pipe by fixture units.

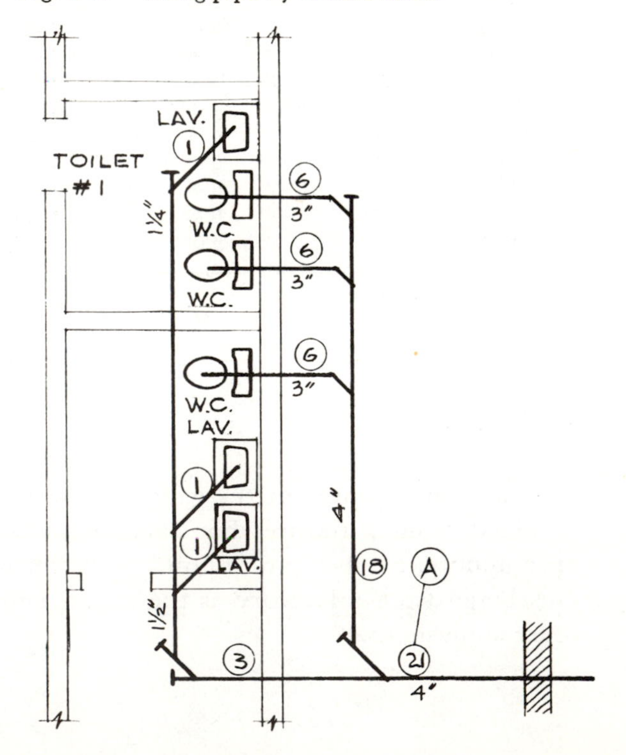

2-15 SIZE OF PIPE FOR SANITARY DRAINAGE

The partial plumbing plan of Fig. 2-17 shows two toilet rooms, one with one lavatory and two water closets, and one toilet room with one water closet and two lavatories. In sizing the pipe we must first assign the proper fixture unit number to each fixture. Refer to Table 2-3.

The fixture units on the piping layout are represented by the circled numbers. The accumulated number of fixture units at A is 21. The pipe size can now be found by matching the fixture units with the pipe size as shown in Table 2-4.

In the first toilet room the lavatory has one fixture unit. One fixture unit calls for a pipe size of 1¼ in. (32 mm). The three fixture units on the lavatory pipe line require a pipe size of 1½ in. (38 mm). For a single water closet of six fixture units, the minimum pipe size is 3 in. (76 mm). With three water closets totalling 18 fixture

TABLE 2-4

Maximum Permissible Loads for Sanitary Drainage Piping (in terms of fixture units)

Pipe Diameter (in.)	Any Horizontal Fixture Branch	One Stack of 3 Stories or Less in Height	Stacks more than 3 Stories in Height		Building Drain, and Building Drain Branches from Stacks			
			Total for Stacks	Total at One Story	Slope (in. per ft) ⅟₁₆	⅛	¼	½
1¼[a]	1	2	2	1	–	–	–	–
1½[a]	3	4	8	2	–	–	–	–
2 [a]	6	10	24	6	–	–	21	26
2½[a]	12	20	42	9	–	–	24	31
3	20[b]	30[c]	60[c]	16[b]	–	20[b]	27[b]	36
4	160	240	500	90	–	180	216	250
5	360	540	1100	200	–	390	480	575
6	–	960	1900	350	–	700	840	1000

[a]No water closet permitted
[b]Not over two water closets permitted
[c]Not over six water closets permitted

units, the pipe size required becomes 4 in. (102 mm). For a total of 21 fixture units the pipe size is still 4 in.

2-16 REVIEW QUESTIONS

1. Name the two types of drainage systems.
2. What portion of pipe is the house sewer?
3. How does a soil stack differ from a waste stack?
4. The piping for the house sewer may be of what material?
5. What is the minimum pitch or slope of the house sewer?
6. Explain the purpose of the house trap.
7. Why must the house trap be vented by a fresh air inlet?
8. Why must branch vent pipes be graded or pitched?
9. What is the minimum diameter of vent pipe?
10. Where must cleanouts be provided?

3 Storm Drainage Systems

3-1 INTRODUCTION

Many cities have or are building separate drainage systems for storm water and separate drainage systems for sanitary purposes. Each system has its own leaders, house drains, house traps, and fresh air inlets. In some areas, especially where tract homes are built, house traps and fresh air inlets are not used.

Where a combination sanitary and storm sewer system is used, such as shown in Fig. 3-1, the storm leader system should be connected to the house sewer outside the building rather than to the house drain, so that when a separate storm system is constructed later, the cost of changing the connection from the combination sewer to the storm water system will be much less than where leaders are connected to the house drain. The house trap and the fresh air inlet (FAI) are not used in many areas of the country.

3-2 GUTTERS AND LEADERS

Water from a pitched roof flows into a roof gutter which may be a half round or square-shaped trough as in Fig. 3-2. Gutters may be made of wood, aluminum, or galvanized sheet metal. Almost

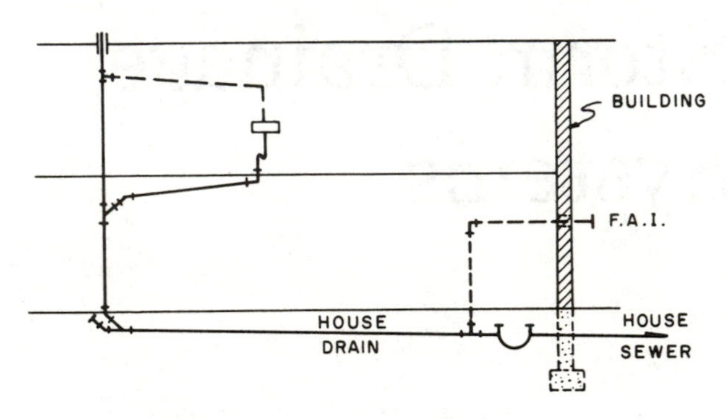

SANITARY SEWER

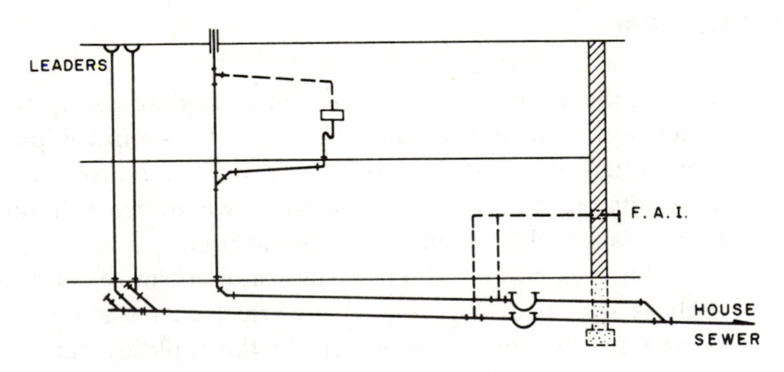

COMBINATION SANITARY & STORM SEWER

Fig. 3-1 Storm and sanitary sewer. House trap and FAI may not be used depending on location of building.

1. Gutter
2. Leader

Fig. 3-2 Gutter and leader.

all houses have gutters except in areas of the country where there is little rainfall.

Table 3-1 gives the sizes of gutters required for the maximum rainfall in inches per hour for the various roof areas shown. The sizes are for gutter slopes of 1/16 in., 1/8 in., 1/4 in. and 1/2 in. per foot of length.

EXAMPLE: Find the size of a semicircular roof gutter (Table 3-1) when the roof area is 1090 sq ft and the maximum rainfall per hour is 5 in. The gutter is set at 1/8-in. slope. Find the roof area in the column under 5-in. rainfall; 1085 sq ft is closest value and thus we read 6-in. diameter gutter in the left hand column.

From the roof gutter, the water flows down a pipe called a leader. The leader carries the water down into the ground and into a drywell, which often consists of an oil drum filled with rocks and punctured with holes for the water to run out and disperse into the ground. Leaders may also stop just before they reach the ground so that the water spills on a splash tray and runs into the surrounding ground area. Metal straps are used to fasten the leader to the side of the house.

TABLE 3-1

Size of Gutters[a]

Diameter of gutter (in.) (1/16 in. slope)	Maximum Rainfall (in. per hr)				
	2	3	4	5	6
3	340	226	170	136	113
4	720	480	360	288	240
5	1,250	834	625	500	416
6	1,920	1,160	960	768	640
7	2,760	1,840	1,380	1,100	918
8	3,980	2,655	1,990	1,590	1,325
10	7,200	4,800	3,600	2,880	2,400

Diameter of gutter (in.) 1/8 in. slope	Maximum Rainfall (in. per hr)				
	2	3	4	5	6
3	480	320	240	192	160
4	1,020	681	510	408	340
5	1,760	1,172	880	704	587
6	2,720	1,815	1,360	1,085	905
7	3,900	2,600	1,950	1,560	1,300
8	5,600	3,740	2,800	2,240	1,870
10	10,200	6,800	5,100	4,080	3,400

Diameter of gutter (in.) 1/4 in. slope	Maximum Rainfall (in. per hr)				
	2	3	4	5	6
3	680	454	340	272	226
4	1,440	960	720	576	480
5	2,500	1,668	1,250	1,000	834
6	3,840	2,560	1,920	1,536	1,280
7	5,520	3,680	2,760	2,205	1,840
8	7,960	5,310	3,980	3,180	2,655
10	14,400	9,600	7,200	5,750	4,800

Diameter of gutter (in.) 1/2 in. slope	Maximum Rainfall (in. per hr)				
	2	3	4	5	6
3	960	640	480	384	320
4	2,040	1,360	1,020	816	680
5	3,540	2,360	1,770	1,415	1,180
6	5,540	3,695	2,770	2,220	1,850
7	7,800	5,200	3,900	3,120	2,600
8	11,200	7,480	5,600	4,480	3,730
10	20,000	13,330	10,000	8,000	6,660

[a]Courtesy of IAPMO (International Association of Plumbing and Mechanical Officials)

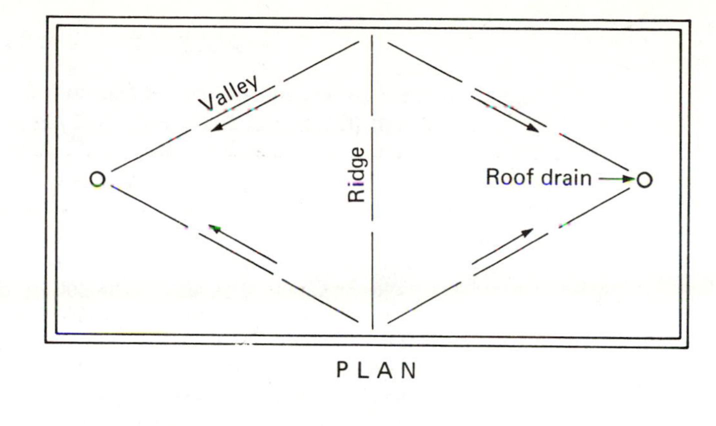

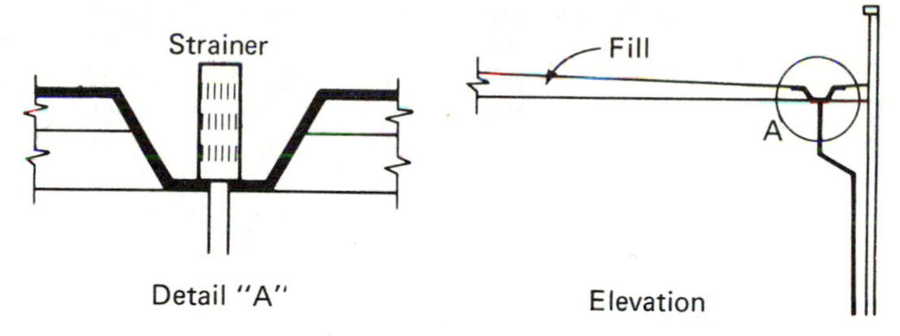

Fig. 3-3 Flat roof drainage.

3-3 FLAT ROOF DRAINAGE

It is easy to see how storm water flows off a pitched roof into a roof gutter and down into the leader. On flat roofs, however, the water drains into roof drains due to a slight pitch created by a roof fill, as seen in Fig. 3-3. The high points are the *ridges* and the low points are the *valleys* that pitch toward the drains. Leaders for flat roofs, such as at A, often run inside of the building, and must, therefore, be made of bell and spigot iron pipe with joints caulked with hemp (oakum) and lead and made watertight. The roof drain is equipped with a strainer to keep out leaves and prevent clogging of the drain. Where the strainer passes through the roof it must be made watertight.

Leaders, or vertical rain water piping, are sized according to Table 3-2. The table is based on maximum inches of rainfall per hour falling on a given roof area in square feet. Local rainfall figures should be consulted to determine maximum rainfall per hour.

TABLE 3-2

Sizing of Roof Drains and Rainwater Piping for Varying Inches of Rainfall (Quantities Given are Roof Areas in Sq Ft)

Rainfall (in. per hr)	Size of Drain or Leader (in.)[a] 2	3	4	5	6	8
1	2,880	8,800	18,400	34,600	54,000	116,000
2	1,440	4,400	9,200	17,300	27,000	58,000
3	960	2,930	6,130	11,530	17,995	38,660
4	720	2,200	4,600	8,650	13,500	29,000
5	575	1,760	3,680	6,920	10,800	23,200
6	480	1,470	3,070	5,765	9,000	19,315
7	410	1,260	2,630	4,945	7,715	16,570
8	360	1,100	2,300	4,325	6,750	14,500
9	320	980	2,045	3,845	6,000	12,890
10	290	880	1,840	3,460	5,400	11,600
11	260	800	1,675	3,145	4,910	10,545
12	240	730	1,530	2,880	4,500	9,660

[a]Round, square, or rectangular rainwater pipe may be used and are considered equivalent when enclosing a scribed circle equivalent to the leader diameter. Courtesy of IAPMO (International Association of Plumbing and Mechanical Officials)

3-4 DESIGN OF STORM WATER SYSTEMS

Sizing Vertical Pipes or Leaders

The flat roof of the small manufacturing plant shown in Fig. 3-4 has an area of 4640 sq ft. Each corner of the roof is a drain or leader, marked L-1, L-3, L-4, and L-6 respectively. A smaller roof over a loading platform has an area of 128 sq ft, with one leader marked L-2, while a front canopy roof of 116 sq ft has one leader marked L-5. Two yard drains are designed to drain an equal number of square feet.

Let us assume the building is located in an area where the rainfall is 4 in. per hour. Leaders L-1, L-3, L-4, and L-6 all carry an equal number of square feet of rainfall, which is 4640 ÷ 4 or 1160 sq ft. By referring to Table 3-2 opposite a rainfall of 4 in., the nearest higher square feet to 1160 is 2200, which requires a 3-in. pipe. Similarly, leaders L-3, L-4, and L-6, having the same

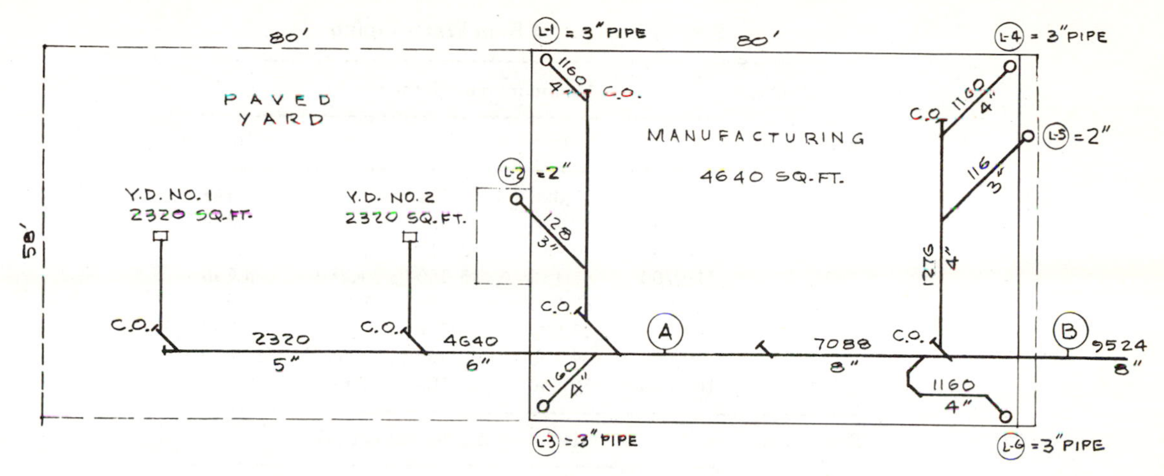

Fig. 3-4 Plan of storm drainage.

number of square feet will also require a leader size of 3 in. Leader L-2 carries 128 sq ft, which requires a 2-in. minimum pipe, and leader L-5, draining 116 sq ft will require also a 2-in. minimum pipe.

Sizing Horizontal Rain Water Piping

The size of the horizontal rain water piping and its branches are sized according to Table 3-3, based upon maximum roof areas to be drained. For example, in our manufacturing plant the horizontal piping from leader L-1 drains 1160 sq ft. A rainfall of 4 in., and a ⅛-in. pipe pitch per foot, the nearest square feet in Table 3-3 is 1880 which calls for a 4-in. pipe. The horizontal pipe from leader L-2 drains 128 sq ft which requires a minimum pipe size of 3 in. Yard drain 1 drains 2320 sq ft of surface rain water. According to Table 3-3, its pipe size must be 5 in. Where both yard drains combine the pipe carries 4640 sq ft of rain water which requires a pipe size of 6 in. At point A, for example, the pipe size is found by adding the two yard drains, plus leader L-1, L-2 and L-3 which equals 7088 sq ft of rain water which demands a pipe size of 8 in. Similarly, add the square feet of rain water carried in a branch pipe to that found before.

At point B, for example, the square feet of rain water for all piping is added making a total of 9524 sq ft, which calls for a 8-in. diameter pipe.

TABLE 3-3

Size of Horizontal Rain Water Piping

Size of pipe (in.) 1/8-in. pitch	Maximum rainfall (in. per hr)				
	2	3	4	5	6
3	1,644	1,096	822	657	548
4	3,760	2,506	1,880	1,504	1,253
5	6,680	4,453	3,340	2,672	2,227
6	10,700	7,133	5,350	4,280	3,566
8	23,000	15,333	11,500	9,200	7,600
10	41,400	27,600	20,700	16,580	13,800
11	66,600	44,400	33,300	26,650	22,200
15	109,000	72,800	59,500	47,600	39,650

Size of pipe (in.) 1/4-in. pitch	Maximum rainfall (in. per hr)				
	2	3	4	5	6
3	2,320	1,546	1,160	928	773
4	5,300	3,533	2,650	2,120	1,766
5	9,440	6,293	4,720	3,776	3,146
6	15,100	10,066	7,550	6,040	5,033
8	32,600	21,733	16,300	13,040	10,866
10	58,400	38,950	29,200	23,350	19,450
12	94,000	62,600	47,000	37,600	31,350
15	168,000	112,000	84,000	67,250	56,000

Size of pipe (in.) 1/2-in. pitch	Maximum rainfall (in. per hr)				
	2	3	4	5	6
3	3,288	2,295	1,644	1,310	1,096
4	7,520	5,010	3,760	3,010	2,500
5	13,360	8,900	6,680	5,320	4,450
6	21,400	13,700	10,700	8,580	7,140
8	46,000	30,680	23,000	18,400	15,320
10	82,800	55,200	41,400	33,150	27,600
12	133,200	88,800	66,600	53,200	44,400
15	238,000	158,800	119,000	95,300	79,250

Courtesy of IAPMO (International Association of Plumbing and Mechanical Officials)

3-5 YARD AND AREA DRAINS

The function of a yard and area drain is to carry off storm water as quickly as it falls, thereby preventing the formation of pools of water.

The capacity of a yard and area drain depends on the rate of rainfall. This may be assumed at 4 in. (102 mm) per hour, depending on the locality. Thus an area of 60 ft X 100 ft (18 X 30.5 m) to be drained would yield the following quantity of water:

$$60 \times 100 = 6000 \times \frac{4}{12} = \frac{24000}{12} = 2000 \text{ cu ft/hr}$$

$$2000 \div 60 \text{ (min)} = 33.3 \text{ cu ft per min.}$$

$$33.3 \times 7.5 = 249.75 \text{ or } 250 \text{ gal per min}$$

Note: 1 cu ft = 7.5 gal

In paved yards, such as courts, schoolyards, and other similar places, catch basins (Fig. 3-5) are built below the ground surface with their metal grating covers level with the finished grade. Catch basins may be of brick or concrete. All paved areas, yards, courts,

Fig. 3-5 Catch basin.

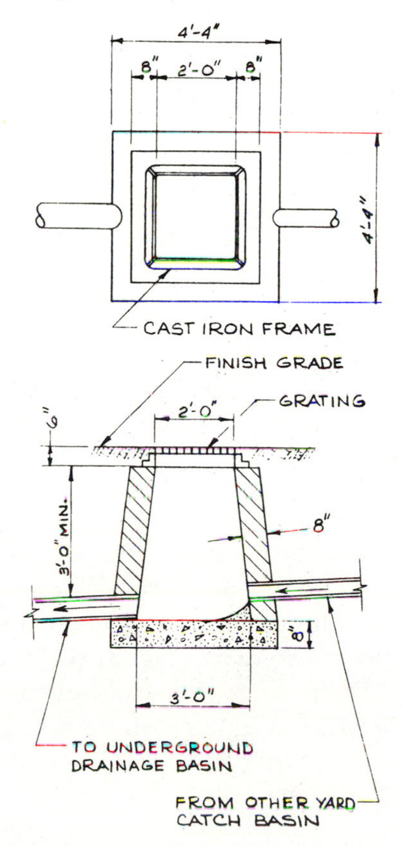

etc. exceeding 15 sq ft (1.39m^2) should be drained into the storm water sewer system, if such is available, by first entering a catch basin and then running in pipes to a storm water sewer system or to an underground drainage basin.

3-6 UNDERGROUND DRAINAGE BASINS

The type of basin shown in Fig. 3-6 is manufactured in three sections: drain ring, solid ring, and dome. Underground basins are used singly or in series to collect rain water from paved areas. The water either runs into the opening at the top of the dome or it is collected in a series of yard-drain inlets, or catch basins, which are connected by pipes to the underground drainage basin.

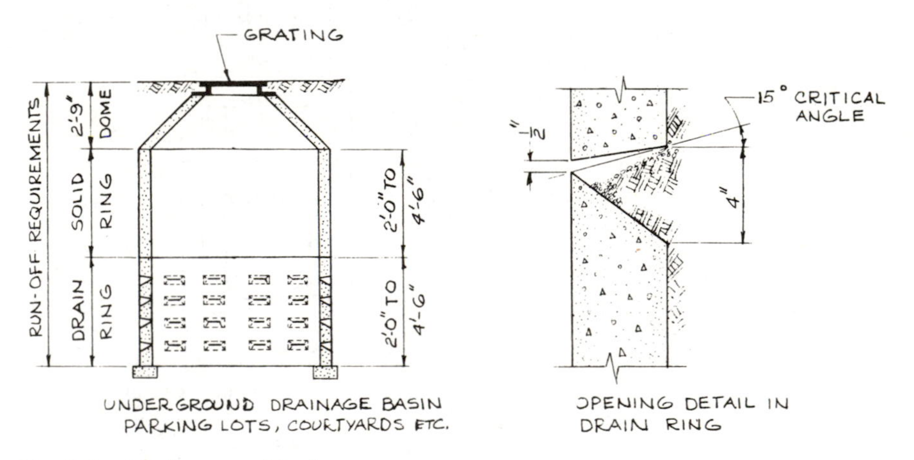

Fig. 3-6 Underground basin.

The lower ring of the underground basin is provided with drain hole openings designed especially to let the water seep into the earth without allowing the openings to become clogged with earth. Drainage also takes place at the open bottom of the basin.

Drain rings and solid rings are obtainable in 4-, 6-, 8-, and 10-ft (1.2-, 1.8-, 2.4- and 3-m) diameters. The heights of individual rings are either 2 ft (610 mm) or 4 ft 6 in. (1.37 m). They can be stacked as needed.

3-7 DRAINAGE REQUIREMENTS

Standards have been developed establishing site drainage of projects submitted for approval to planning boards and town engineers. These are as follows:

1. All storm water shall be contained on the site.
2. Where soil conditions and all other requirements of the town engineer so justify, an underground basin shall be selected in accordance with the runoff factors of Table 3-4.
3. Computations for cubic footage of rainfall shall be determined as follows: required capacity in cubic feet is equal to the area in square feet to be drained, multiplied by the proper runoff factor, multiplied by 0.12 ft or 1½ in (38 mm), the depth of the rainfall in the particular locations.
4. An underground basin with drainage openings, shall be designed, allowing for ground absorption, as follows:

$$\text{Total capacity} = \text{Area (sq ft)} \times \text{Runoff factor} \times 0.12 \text{ ft}$$

5. Selection of underground basins shall be guided as follows: Standard 8 ft-0 in. (2.4-m) diameter basin contains 42 cu ft (1.19 m^3) for each foot (0.3 m) of height. Standard 10 ft-0 in. (3.0 m) diameter basin contains 67 cu ft (1.89 m^3) for each foot of height.

No credit for drainage shall be given for basin height above the level of inlet pipes.

TABLE 3-4

Runoff Factors

Pavement (grade 5% or less)	0.90
Pavement (grade in excess of 5%)	0.95
Seeded (grade 5% or less)	0.08
Seeded (grade in excess of 5%)	0.17
Roof deck	1.00
Other areas (minimum)	0.30

3-8 A TYPICAL PROBLEM

Find the equivalent drainage for the following developed land area and building.

Area (sq ft)	Use	Material	Runoff Factor	Drainage (sq ft)
6,500	Parking	Paving	0.90	5,850
6,500	Building	Roof deck	1.00	6,500
3,000	Landscaping	Seeded (grade 5% or less)	0.08	240
500	Undeveloped	Earth	0.30	150
16,500	Total	Equivalent Drainage		12,740

Total capacity: 12,740 sq ft X 0.12 ft (rainfall) = 1528.8 cu ft

$$\frac{1528.8 \text{ cu ft}}{42 \text{ cu ft}} = 36.40 \text{ lin ft}$$

Using a 4 ft -6 in high basin of 8 ft -0 in. diameter, divide the (height) linear feet by 4 ft -6 in., or $\frac{36.4}{4.5} = 8.1$ or eight sections of 4 ft -6 in. length. To stack up eight sections of 4 ft - 6 in. lengths in one stack is not practical. It would require a hole in the ground of at least 36 ft deep.

The 36-ft high stack of rings can be broken up and placed into two 18-ft holes, or four 9-ft holes, or eight 4½-ft holes depending on the height of the water table. The bottom of all underground basins must be at least 2 ft above the water for the proper drainage of the basin.

Similarly, in using a 10-ft (3 m) diameter basin which can hold 67 cu ft (1.89 m^3) per foot of height, the linear feet of basin needed will be $\frac{1528.8 \text{ cu ft } (43.27 \text{ m}^3)}{67 \text{ cu ft}} = 22.82$ lin ft (6.96 m).

Using a 4 ft-6 in. (1.37m) high basin section, divide the linear feet by 4 ft-6 in., or $\frac{22.82}{4.5}$ = five sections of 4 ft-6 in. lengths. One stack will require a 22.5-ft hole, which most likely is too deep. It can be broken up into three sections of 4½-ft lengths and two sections of 4½-ft lengths located in different places on the drained area.

3-9 BUILDING YOUR OWN SEEPAGE PIT

Where buildings are provided with public sewers, seepage pits are not necessary. Both water used by appliances and sewage flow into the house drain to the house sewer and then into the public sewer system.

However, in areas where there are no public sewers, both sanitary sewage and waste water from water-using appliances flow into a septic tank, where the solid matter settles to the bottom of the tank and the effluent or liquid part drains into the ground through a system of open-jointed piping laid in the ground.

With an increase in house water-using appliances such as washing machines, multiple shower baths, and bathtubs, an additional burden can be placed on the household drainage system, causing the septic tank to overflow. Washing machines account for about 40 to 45 percent of the average family's waste water.

To overcome this nuisance, a simple and relatively inexpensive solution is possible that allows the home owner to put in some fixtures without taxing, or even tampering with the septic tank. The solution is to build a seepage pit or drywell (Fig. 3-7) which is a separate drainage facility that can dispose of clean waste water that is free of solids or germ laden material and therefore, does not require septic tank treatment.

A seepage pit or drywell is nothing more than a lined hole in the ground that collects water and slowly allows it to disperse into the surrounding soil.

The hole can be lined with stone or unmortared concrete block, specially made for this purpose. The hole should have a precast concrete cover with an access opening and a cover at the top as shown in Fig. 3-8. Other seepage pits are precast concrete cylinders that can be purchased in various diameters and heights.

3-10 TESTING SOIL FOR SEEPAGE

Before digging a seepage pit it may well be worth your while to first test the soil for its rate of water absorption. Dig a hole with a post-hole auger 12 in. (305 mm) in diameter and about 2-ft (610 mm) deep. Should you hit rock or clay move the hole elsewhere. If ordinary earth is encountered after digging 2 ft down, place about 2

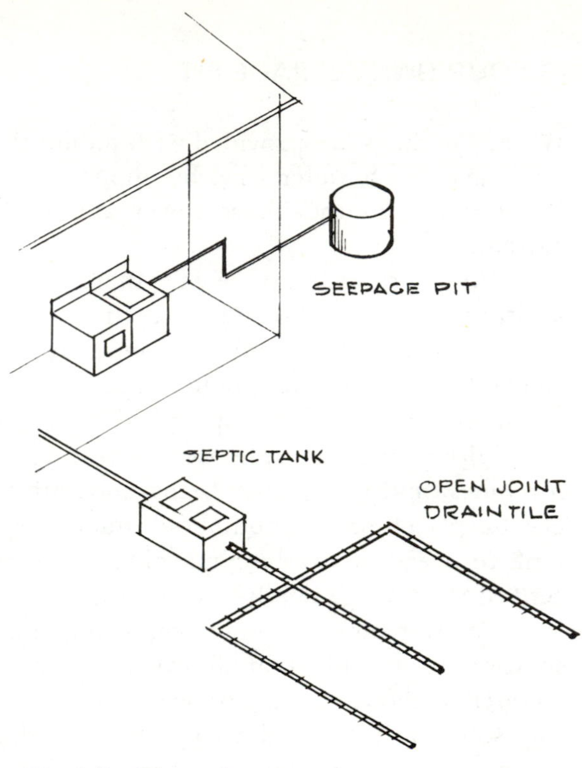

Fig. 3-7 Diverted waste water.

Fig. 3-8 Drywells or seepage pits.

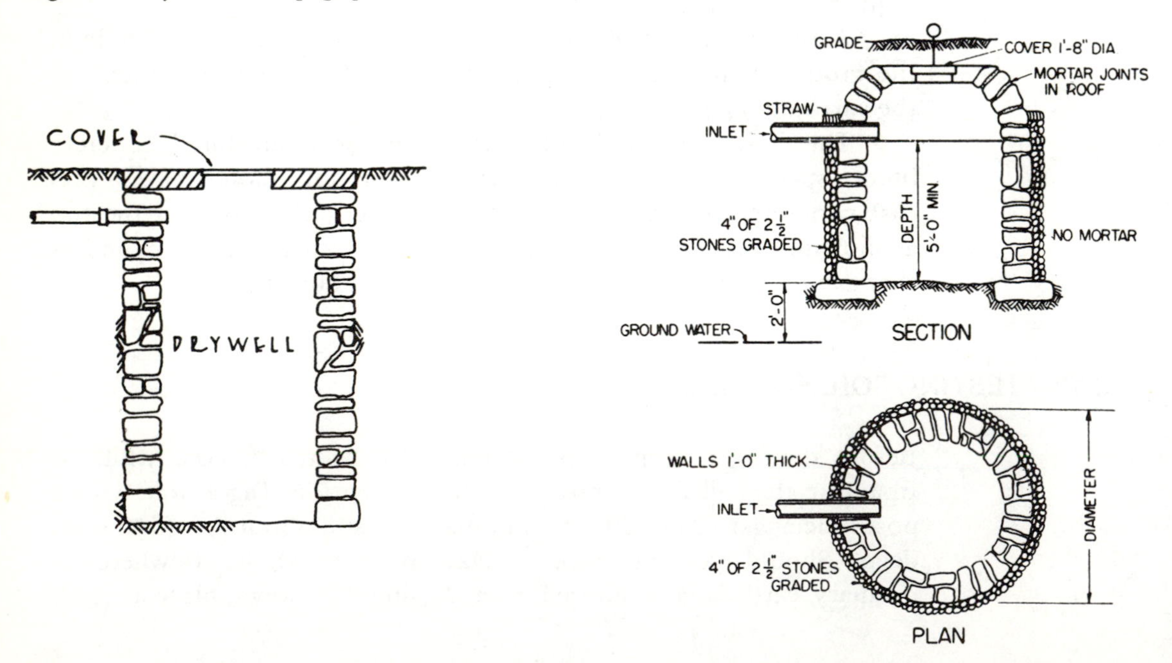

to 3 in. (51 or 76 mm) of gravel in the bottom of the hole and fill the hole with water. Let the water seep away into the surrounding soil. Repeat this a second time. The soil will be saturated with water and therefore ready for testing.

3-11 RATE OF WATER ABSORPTION

The size of a seepage pit depends largely on the rate of the water absorption of the soil. To determine this rate, fill the test hole with water and place a floating wooden disk with a dowel stick marked in inches on the surface of the water. Place a reference board at the top of the hole and note the lowering of the dowel as the water drops as shown in Fig. 3-9. Record the number of minutes the water takes to drop 1 in. (25.4 mm). Wait 30 min and repeat the test. Repeat again until you get the same results two successive times. This will give you the rate of percolation, which may range from a few seconds to 30 min or more. Anything higher than 30 min shows that the seepage or a pit in this location is unsuitable. If the soil is suitable, pit dimensions for various percolation rates are as shown in Table 3-5.

Fig. 3-9 Measuring percolation rate.

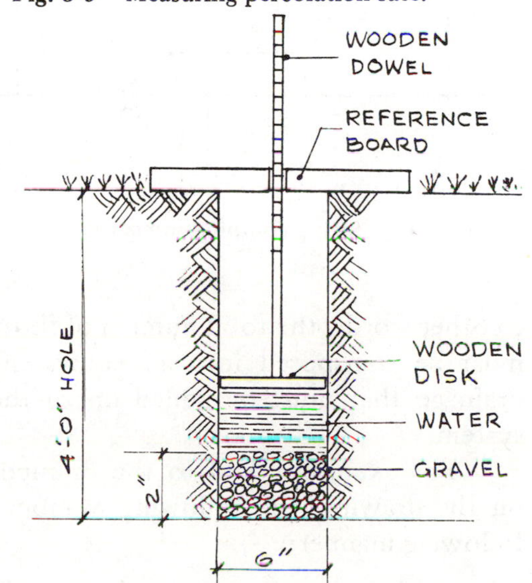

TABLE 3-5

Seepage Pit Dimensions

Percolation Rate (min per in.)	Pit Dimensions
0.5	4 ft deep X 5 ft diameter
6 - 10	4 ft deep X 7 ft diameter
11 - 15	4 ft deep X 9 ft diameter
16 - 20	5 ft deep X 8½ ft diameter

3-12 COMBINED STORM AND SANITARY DRAINAGE SYSTEMS

To find the diameter of the pipe where the sanitary and the storm sewer pipes join (Fig. 3-10), add to the storm drained area an allowance in square feet for each fixture unit on the sanitary system.

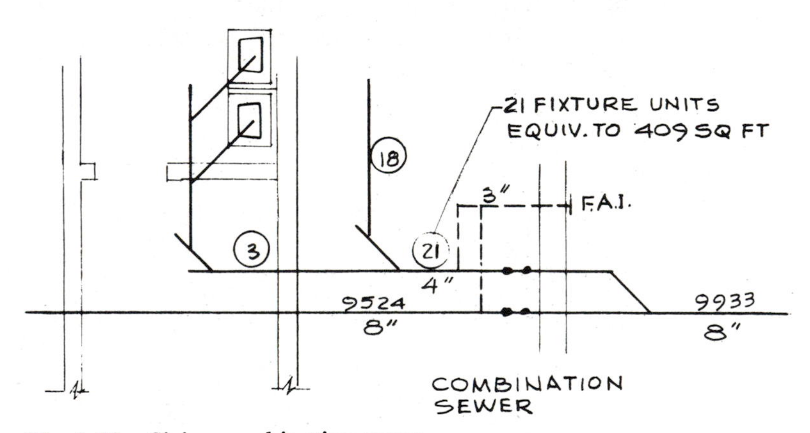

Fig. 3-10 Sizing combination sewer.

In other words, the total number of fixture units (21 in our example), must be transposed into an equivalent number of square feet of drainage that must be added on to the drained area of the storm system.

For example: Add to the drained area of 9,524 sq ft, shown on the drawing, the following number of square feet, found in the following manner:

30 sq ft for each of the first 6 fixture units
30 sq ft for each of the next 4 fixture units
14 sq ft for each of the next 10 fixture units
9 sq ft for each of the next 10 fixture units

$$
\begin{array}{r}
\text{Total fixture units} = 21 \\
-\ 6 \\
\hline
15 \\
-\ 4 \\
\hline
11 \\
-10 \\
\hline
1
\end{array}
$$

Therefore:

$$
\begin{array}{rcrcr}
30 & \times & 6 & = & 180 \text{ sq ft} \\
20 & \times & 4 & = & 80 \text{ sq ft} \\
14 & \times & 10 & = & 140 \text{ sq ft} \\
9 & \times & 1 & = & 9 \text{ sq ft} \\
\hline
 & & & & 409 \text{ sq ft}
\end{array}
$$

Add the 409 sq ft to the 9,524 sq ft of the storm system; a total of 9,933 sq ft.

The additional equivalent square feet has not changed the pipe diameter requirement of the combination sewer. See Table 3-3, for a rainfall of 4 in. and pipe pitch of ⅛-in. per foot. The nearest largest square feet is 11,500 making the required pipe size 8 in. in diameter.

3-13 BACKWATER PREVENTION FOR ORDINARY CONDITIONS

Overloading of the city sewer often results in the flooding of basements. Figure 3-11 shows how the basement may be entirely cut off from drain connections through which backwater may enter. It would be necessary for the water in the sewer (and in the catch basin) to reach the height of the laundry tub rim to become a hazard. The catch basin is shown since certain localities make it mandatory. The catch basin acts as a large grease trap, allowing only clear water to enter the sewer. It should be noted that fixtures discharging solid

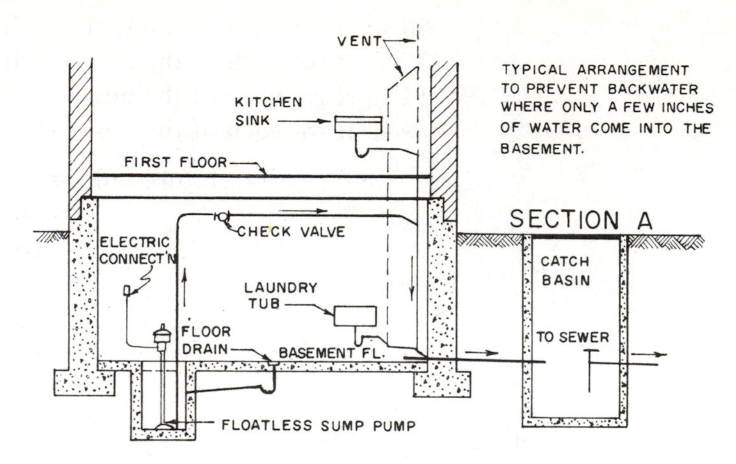

Fig. 3-11

wastes are not drained to the sump pit or to the catch basin. Therefore, they must be above the level of the sewer and connected directly to it through the usual house drain trap.

3-14 BACKWATER PREVENTION FOR SEVERE CONDITIONS

By also draining the laundry tubs into the sump pit as in Fig. 3-12, through a lint trap, the height that is safe against backwater becomes sufficient to provide for the most severe conditions. Otherwise

Fig. 3-12

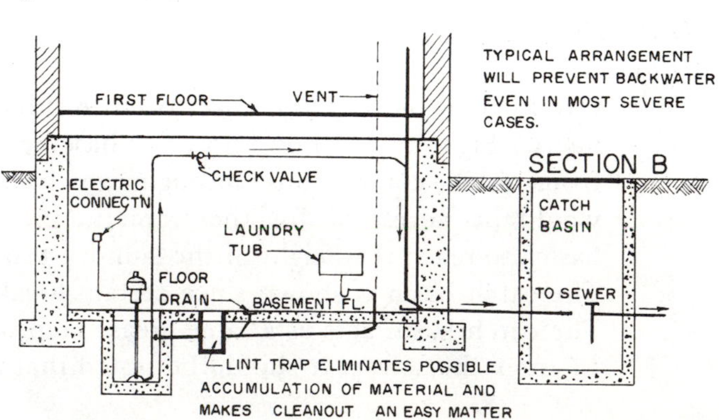

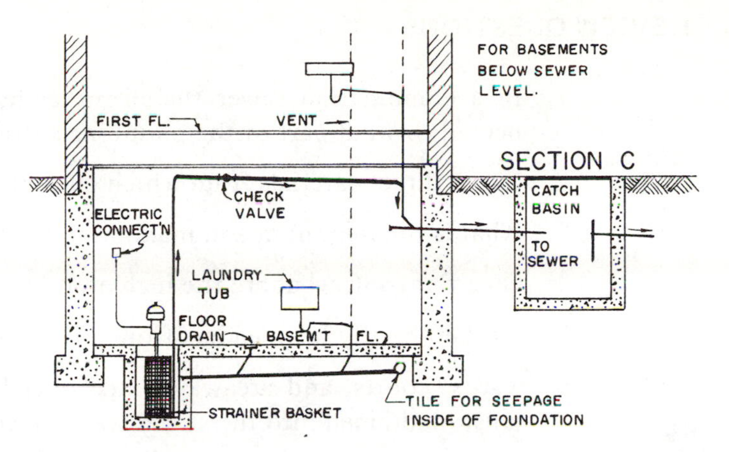

Fig. 3-13

section B corresponds to section A in Fig. 3-11. The lint trap prevents the accumulation of material which might clog the sump pump strainer. See also section C of Fig. 3-13.

3-15 DRAINAGE FOR FIXTURES BELOW SEWER LEVEL

Because the basement floor is considerably below the level of the city sewer does not mean that the occupant must be denied the conveniences of plumbing in the basement. The floor drain for convenience in cleaning laundry tubs and other fixtures not discharging solid waste may be utilized in connection with a floatless sump pump.

Figure 3-13, section C, shows such a piping arrangement. The backing up of the sewer due to storms and other causes cannot flood a basement having such an arrangement of waste lines.

The catch basin is required in many localities to prevent grease and soap from sinks and tubs from reaching the sewer. This illustration shows an alternate method of preventing lint from entering the sump pit, through the use of a strainer basket.

Drainage piping serving fixtures below the elevation of the curb or property line is required to drain by gravity into the main sewer, and must be protected from backflow of sewage by installing a backwater pressure valve (BWPV).

3-16 REVIEW QUESTIONS

1. In a combination sewer the storm leader system must be connected to what part of the plumbing system?
2. Name three materials from which gutters are made.
3. What is the minimum and maximum gutter slope?
4. On a flat roof, what are the high and low points called?
5. Explain briefly how leaders from a flat roof are sized.
6. Yards, courts, and areaways exceeding how many square feet must be drained into the storm water sewer system?
7. Drain rings for underground basins are available in what diameters?
8. In what areas are seepage pits most likely to be built?
9. Explain briefly, how a combination sewer is sized.
10. Explain backwater prevention.

4 Lawn Sprinkler Systems

4-1 HOW TO INSTALL A LAWN SPRINKLER SYSTEM

Here is a simplified step by step procedure for installing a new lawn sprinkler system. Follow the steps carefully and work at your own pace.

STEP 1: Find out the following:

a. Water pressure (in psi) (call the local water department or use pressure gauge like the one shown in Fig. 1-1)
b. Plumbing regulations, building permits needed (call the city hall or building department)
c. Inside diameter of pipe (service line) running from water meter to house (wrap a piece of string around the outside of the service line. Measure the length of the string required to circle it.) Refer to Table 4-1 for service line size.

STEP 2: Use a sheet of graph paper similar to that shown in Fig. 4-1, where each 1 in. square is divided into 10 equal spaces, and make a scale layout of your property. The scale on paper is 1 in. =

TABLE 4-1

Required Pipe Diameter

Length of string	in.	2¾	3¼	3½	4	4⅜	5
	mm	70	82.35	88.7	102	111.5	127
Size of copper tubing	in.	¾		1		1¼	
	mm	19		25.4		31.75	
Size of galvanized pipe line	in.		¾		1		1¼
	mm		19		25.4		31.75

10 ft. Purchase a sheet that is 18 by 20 large squares, if you can. Otherwise tape several smaller sheets together.

a. Include the location of house, driveways, walkways, paths, fences, walls, structures, planters, patios, flower beds, shrubbery, and lawn areas.

b. Locate all areas to be covered by water: lawns, shrubbery, special grounds, etc. A typical residential plan is shown in Fig. 4-2.

STEP 3: Select and position sprinkler heads.

a. Table 4-2 identifies the major types of sprinkler heads, their uses, patterns available, and suggests correct spacing requirements.

b. Begin graphing large lawn areas. Place quarter circle heads in the corners of all perimeter areas. Place half circle heads between corners, following the recommended equal distance spacings. Place full circle heads within each perimeter; the number will depend on the perimeter's overall dimensions. Study Fig. 4-3, a typical yard area showing positioned sprinklers.

c. Now position heads in small lawn areas, pathways, and at the side of the house. These areas are usually watered by one or two rows of part circle heads. See Fig. 4-4.

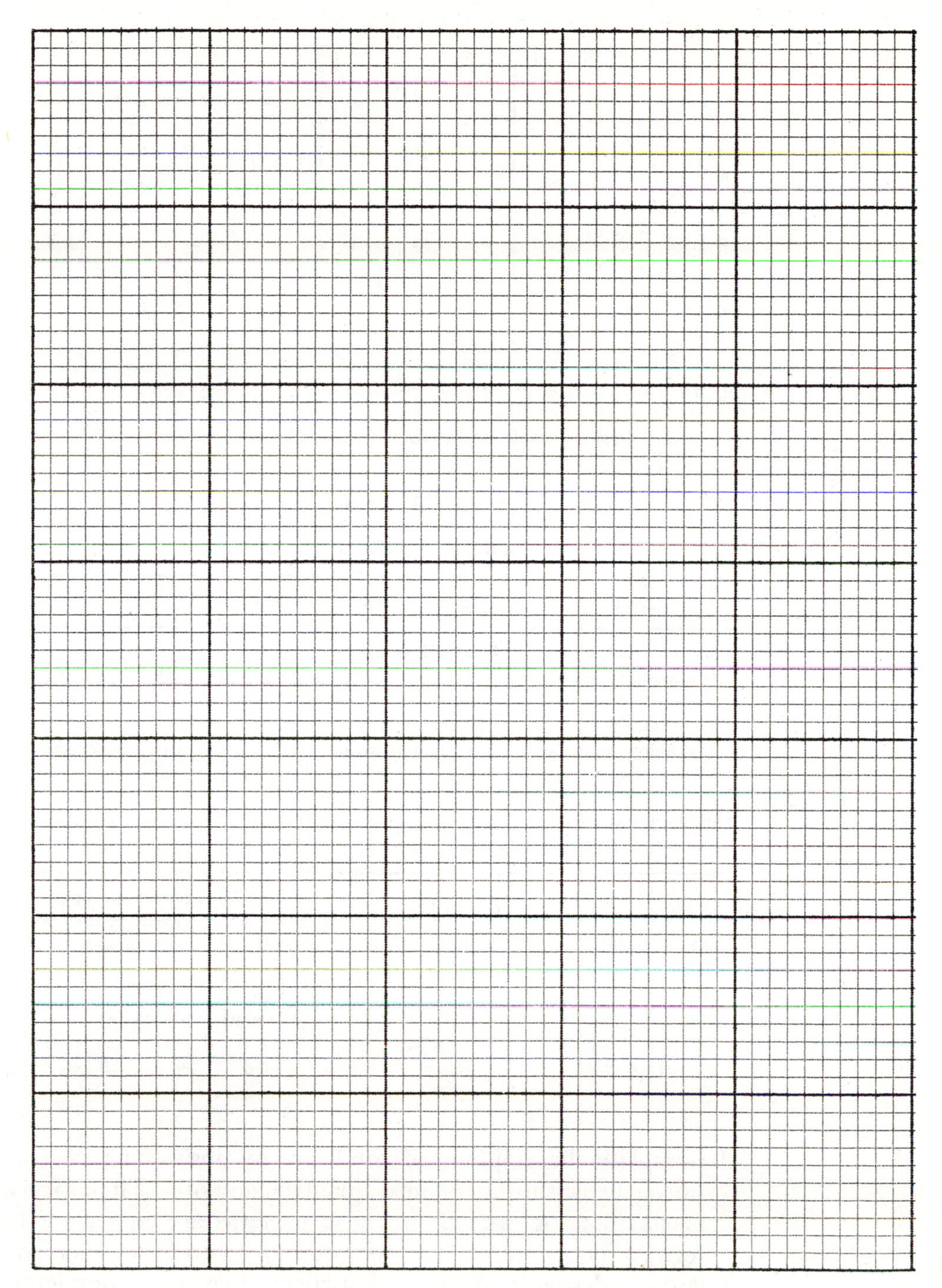

Fig. 4-1

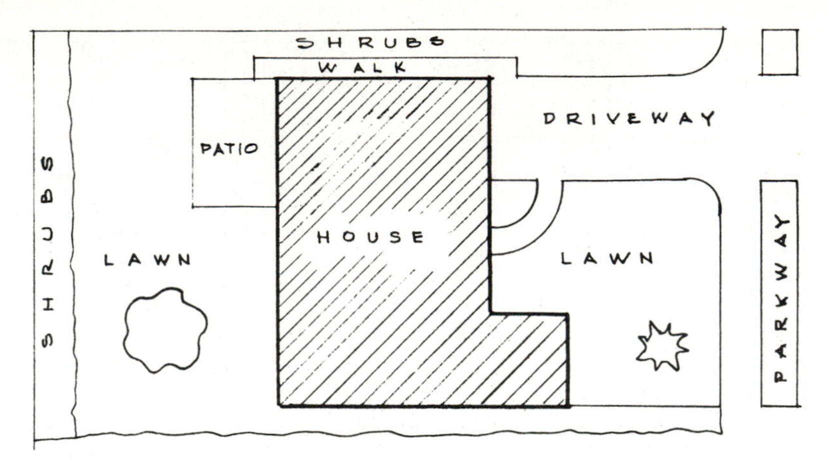

Fig. 4-2 Typical residential plan.

TABLE 4-2

Major Types of Sprinkler Heads

Sprinkler Type	Spray Patterns	Spacing	Uses
Flush sprinkler with flow adjustment	Full, half, quarter, three quarter, strip	12 ft (3.65 m)	Light lawn area
Pop-up sprinkler with flow adjustment	Full, half, quarter, three quarter, strip	12 ft (3.65 m)	Heavy lawn area
Shrubbery sprinkler with flow adjustment	Full, half, quarter, three quarter, strip	12 ft (3.65 m)	Shrubs, ground covers, covering large areas
Bubbler sprinkler	Full circle bubbling	Space as needed	Shrubs, ground covers, plants covering small area
Flooding sprinkler	Full circle, six side jets	18 in (457 mm) Space as needed	Shurbs, ground covers, plants covering small area

d. Complete the graphing of the heads by positioning them in flower beds, shrubbery areas, planters, and other large special places. See Fig. 4-5. Shrub head patterns are equivalent to regular head patterns. Some overlap onto sidewalks and curbs is desirable to ensure that edges of lawn are completely watered.

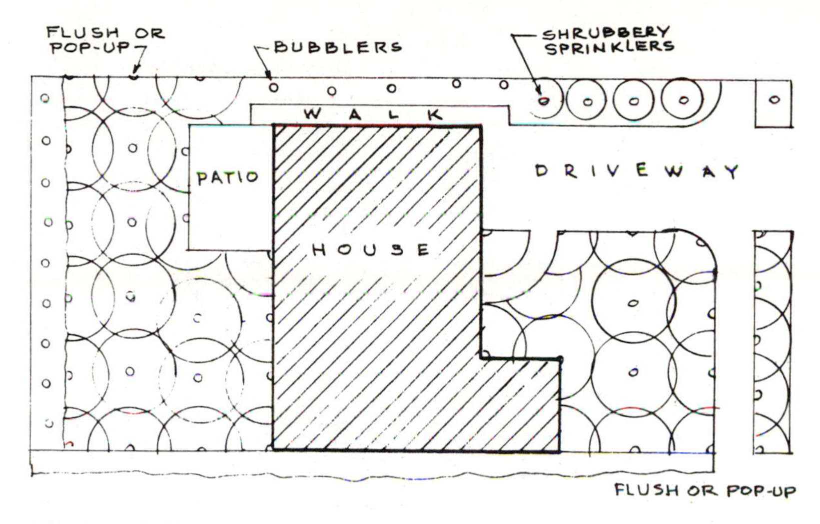

Fig. 4-3 Positioning of sprinklers.

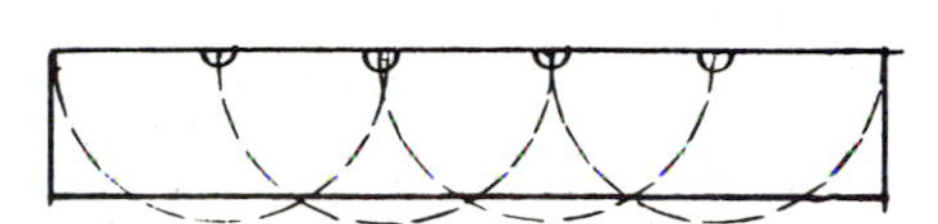

Fig. 4-4 Part circle heads.

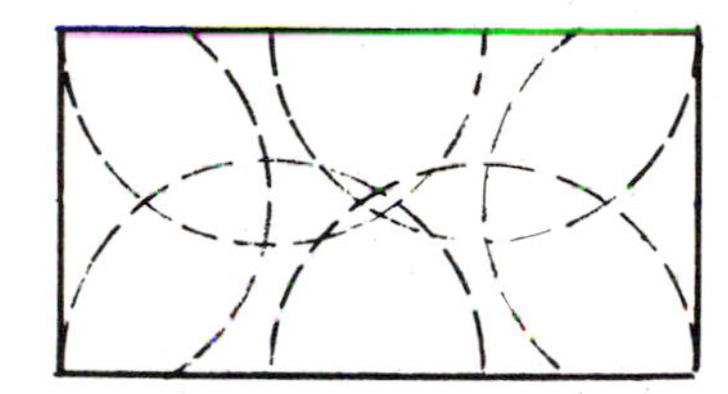

Fig. 4-5 Shrub-head pattern.

e. Irrigation heads and bubblers should be used in tree wells, planters, for ground cover, and around plants that require soaking.

STEP 4: Determine the size, number, type, and location of control valves.

a. Valve sizes are determined by the size of the water meter and the size of the service line (Step 1). Valve sizes cannot exceed either the water meter size or the service line size.

b. The number of valves needed is determined by the total number of heads that can operate off a valve of a particular size. See Table 4-3. Divide the total number of heads needed by the appropriate figure from Table 4-3 to determine the number of valves required. For best results, only one valve should be operated at one time. For example, if there is

TABLE 4-3

Total Heads per Valve

Valve Sizes		Standard Water Pressure				Irrigation Water Pressure		
	psi kPa	30 207	40 276	50 345	60 414	20 138	30 207	40 276
¾ in. 19 mm		6	7	8	9	16	22	24
1 in. 25.4 mm		11	13	14	16	24	32	34
1¼ in. 31.75 mm		17	20	23	25			
1½ in. 38.1 mm		25	28	31	34			
2 in. 50 mm		36	40	43	46			

a ¾-in. (19 mm) supply, 40 psi (276 kPa) pressure, and 21 sprinklers are used, three ¾-in. (19 mm) valves will be needed, since 21 ÷ 7 is 3.

c. The type of valve used depends on local regulations. Some cities require an antisyphon control valve to prevent dirt and foreign matter from draining into the main water system. If those are not required, use standard angle valves. For buried valves, use valve sleeves to protect the valve and stem from dirt. Figure 4-6 shows typical valve installations.

Fig. 4-6 Valves.

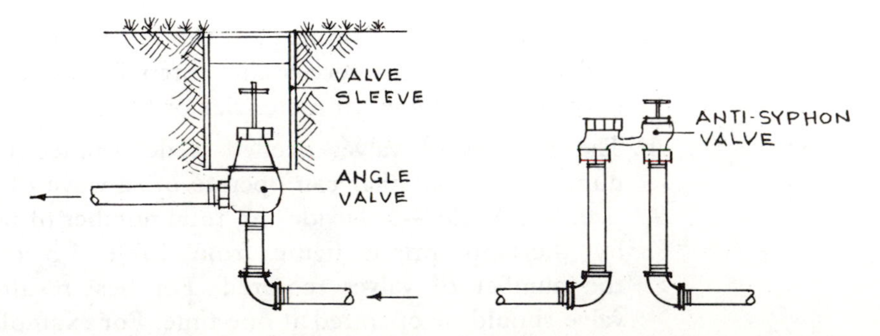

d. Valve locations are determined largely by their nearness to the service line and existing outlets. Pick a convenient location away from heavy foot traffic. Group valves together to save on supply line pipe and other materials.

STEP 5: Determine supply line size.

a. The supply lines are the connecting piping between control valves and either the meter, service line, or existing outlet.

b. Because of friction loss in the supply line, it is necessary to consider the distance from the connection to the valve location. If the tie-in point to the meter, service line, or existing outlet is within 50 ft (15 m) of the valve location, use a supply line pipe equal to the size of the largest control valve; if between 50 and 100 ft (15 and 30 m), use one size larger than the largest control valve; if between 100 and 200 ft (30 and 60 m), use two sizes larger than the largest control valve. If an existing outlet is used, it must be as large as the control valve.

The situation may arise where the supply line is larger than the control valve. In this case it is necessary to reduce the supply line with reducers to the size of the valve.

STEP 6: Group the sprinklers and graph the piping plan.

a. A sprinkler group is the number of sprinkler heads adjacent to each other that will work off of the same control valve. If at all possible, all sprinklers in a group should have the same function. For example, only bubblers should be grouped together. Start at the end of each particular group and work towards its control valve, noting the number of planned heads. If you have three valves, there will be three groups of sprinklers. For example, Table 4-3 shows that 40 psi (276 kPa) pressure with a ¾-in. (19-mm) control valve can handle seven sprinkler heads in a group. The dotted lines on Fig. 4-7 indicate the grouping of sprinkler heads.

b. Start at the control valve and lay out the piping system. Draw a feed line through the middle of each sprinkler group. Draw in side branches to adjacent heads. See Fig. 4-8 for a typical piping layout.

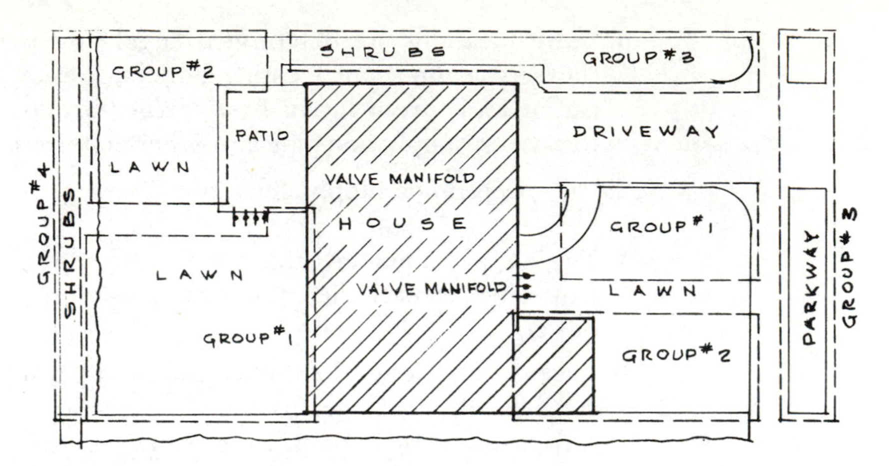

Fig. 4-7 Grouping of sprinklers.

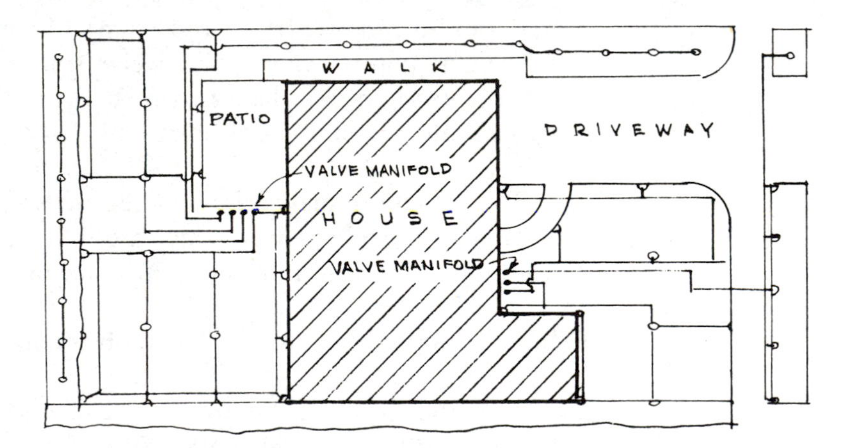

Fig. 4-8 Piping layout.

STEP 7: Review the plan and order what is needed.

a. Check over the details of the plan carefully before ordering sprinkler heads and valves from your nearest retail distributor.

b. Use the convenient Material Required form of Fig. 4-9 to list the items that are needed.

4-2 SET THE CONTROL VALVES

To connect the supply line, shut off the water at the meter, cut a section of the service line and insert a slip coupling tee. Run the supply line to the location of the valves. To connect to an existing

MATERIALS REQUIRED

Description	Quan.	Price	Total
HEADS			
Flush Sprinkler			
Model 17S Full			
Model 17S Half			
Model 17S Quarter			
Other__________			
Pop-Up Sprinkler			
Model 18S Full			
Model 18S Half			
Model 18S Quarter			
Other__________			
Shrubbery Sprinkler			
Model 19S Full			
Model 19S Half			
Model 19S Quarter			
Other__________			
Irrigation Sprinklers			
Model 19-1 Full			
Model 19-2 Full			
Other__________			
Model #29—Sprinkler Head Wrench			
VALVES			
Angle Control Valves			
Model 200—¾"—less Union			
Model 200— 1"—less Union			
Model 300—¾"—with Union			
Model 300— 1"—with Union			
Anti-Syphon Valves			
Model 466—¾"—with Union			
Model 466— 1"—with Union			
Model 466P—¾"—less Union To be used with Plastic Pipe			
Model DV—Drain Valve To be used in areas where pipes can freeze.			
Model #30—Sprinkler Control Key			
Model #20—Valve Sleeve			
PLASTIC PIPE			
Pipe Lengths—¾"x10'			
Pipe Lengths— 1"x10'			
Polyethylene Pipe—¾"x100'			
Polyethylene Pipe— 1"x100'			

Description	Quan.	Price	Total
PLASTIC FITTINGS—Poly or PVC			
Couplings—¾"			
Couplings— 1"			
Tees—SxSxS—¾" x ¾" x ¾"			
Tees—SxSxS—1" x 1" x 1"			
Riser Tees—SxSxT—¾" x ¾" x ½"			
Riser Tees—SxSxT—1" x 1" x ½"			
90° Ells—SxS—¾" x ¾"			
90° Ells—SxS—1" x 1"			
90° Riser Ells—SxT—¾" x ½"			
90° Riser Ells—SxT—1" x ½"			
Male Adapter—¾"—ThreadxSlip			
Male Adapter— 1"—ThreadxSlip			
Plastic Nipple—½" x 6" Poly Cut Off Riser			
Stainless Steel Clamp—¾"			
Stainless Steel Clamp—1"			
Saddle Tee—¾" x ½"			
Saddle Tee—1" x ½"			
Saddle Tee—¾" x ¾"			
Saddle Tee—1" x 1"			
PVC Solvent—1 pt.			
GALVANIZED			
Pipe—1" x 20'			
Pipe—¾" x 20'			
Pipe—¾" x 12'			
Tee—¾" x ¾" x ¾"			
Tee—1" x 1" x 1"			
90° Ell—¾"			
90° Ell—¾" x ½"			
90° Ell—1"			
Coupling—¾"			
Coupling—1"			
Reducer—1" To ¾"			
Close Nipple—¾"			
Manifold Riser—¾" x 12"			
Manifold Riser—1" x 12"			
Nipple—½" x 4"			
Nipple—½" x 5"			
Nipple—½" x 6"			
Nipple—½" x 12"			
Manifold Nipple—¾" x 5"			
Manifold Nipple—1" x 5"			

Fig. 4-9 Materials required. (Courtesy of Champion Sprinkler Equipment.)

garden hose outlet, simply remove the valve and insert a nipple and a tee. Replace the hose valve and run pipe down to the sprinkler valves. See Fig. 4-10.

A group of valves is called a manifold hookup. An example of a manifold hookup for three valves is shown in Fig. 4-11. In a manifold, use fittings and pipe the same size as the valves. If more valves are needed, additional fittings would be added. Always install a master shut off valve ahead of the control valve manifold.

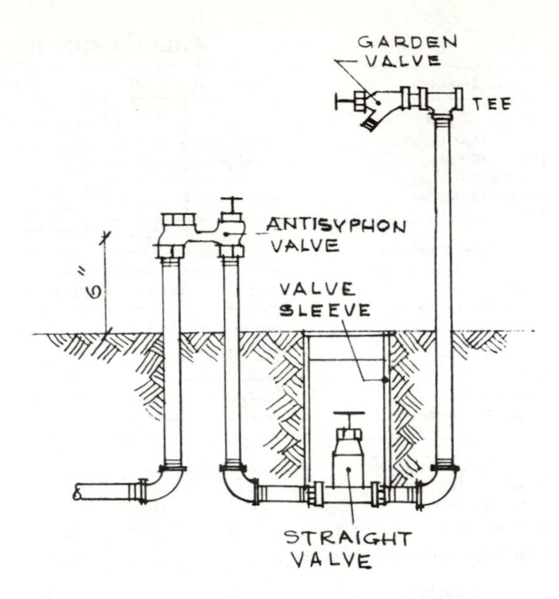

Fig. 4-10 Install tee-run pipe to sprinkler valves.

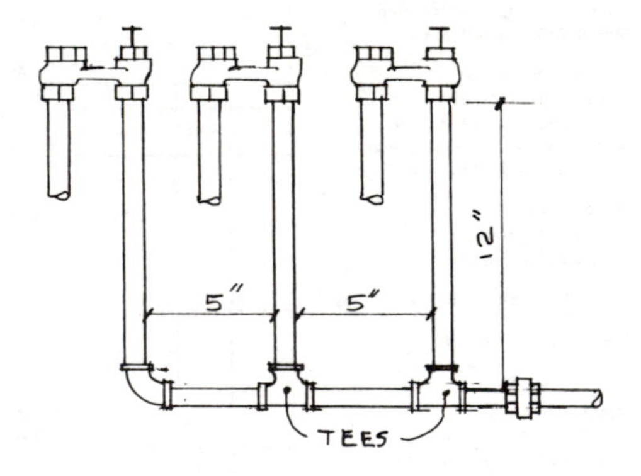

Fig. 4-11 Manifold hookup.

Install valves, making certain that anti-syphon valves are 6 in. (152 mm) above the ground, remembering that angle valves should be buried. Refer again to Fig. 4-6 for typical valve installations for single valve or refer to manifold hookup.

4-3 POSITION SPRINKLER HEADS

Using stakes (for sprinklers) and string (for piping), mark the locations of the heads and piping.

Dig trenches 4- to 5-in. (102- to 127-mm) deep, putting any sod on one side and any dirt or soil on the other side of the trench. For cold climates, lay the pipe so it will drain to a waste valve at its lowest point. Pitch should be about ⅛ in. (3.2 mm) per foot. Hard ground should be watered two days before trenching.

To trench under a cement or brick walk attach a garden hose to the end of a piece of galvanized pipe with a hose-to-pipe adapter. Place the other end of the pipe where you want to trench and turn the water on. Push the pipe under the cement to form the trench.

After hooking up, close all the control valves of the manifold and open the master valve. Check all connections for leaks.

4-4 HOOK UP PIPING

After all trenching is finished, connect all threaded fittings first, starting at the water source. Then connect all slip or insert fittings. If using galvanized pipe, use an adequate amount of pipe joint compound; if using plastic pipe, be certain of a leakproof seal.

How to cut and connect PVC (polyvinyl chloride) pipe:

1. Cut PVC pipe to desired length with a hacksaw.
2. Apply solvent to both pipe and fittings.
3. Push pipe as far as possible into fittings.
4. Wipe off excess solvent.

Using polyethylene pipe:

1. Cut pipe with hacksaw.
2. Put a stainless steel clamp on each section of pipe.
3. Push insert into cut.
4. Tighten clamp over pipe and fitting.

Install sprinkler risers (not heads) as follows: Risers should be within 1 in. (25.4 mm) of the surface for established lawns and should extend a few inches above the ground on temporary risers for new lawns. See Fig. 4-12. Temporary risers should be replaced after the lawn is established and the ground settled. If plastic cut-off risers are used, they can be cut to their permanent length after the lawn is established.

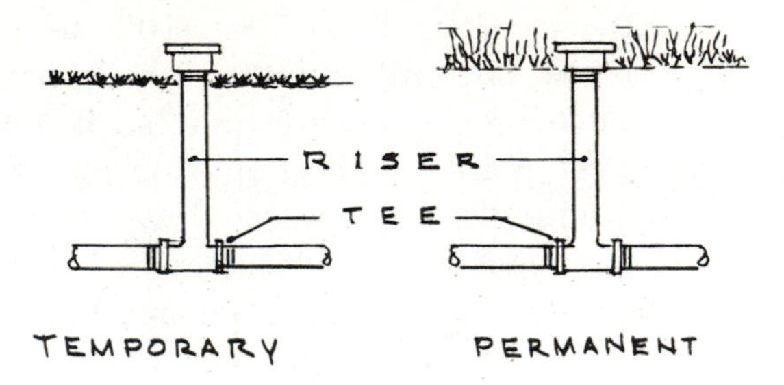

Fig. 4-12 Install sprinkler risers.

4-5 FLUSH OUT THE ENTIRE SYSTEM

Before connecting sprinkler heads, check the final connection of all piping and make certain all dirt and foreign matter are flushed from the piping. Flush out each group individually.

4-6 INSTALL THE HEADS AND CHECK OUT THE SYSTEM

Screw the heads on the ends of riser pipe using a head wrench. Use a screwdriver to turn the adjusting screw and set the spray for correct water coverage. Begin with heads nearest to the valves.

Open each control valve individually and check for leaks and proper coverage.

Fill in open trenches and replace lawn sod or plant new grass.

4-7 HINTS ON OPERATING THE SYSTEM

1. Operate only one valve at a time.
2. Water when pressure will be the greatest. This will most likely be in the morning or late at night.
3. Water shrubs, trees, and plants deeply and less frequently.
4. Water lawn areas for shorter periods and more frequently.
5. Do not water in the heat of the day to cut down on water evaporation. This also will eliminate the burning of plants and lawn areas.

6. Sprinklers should be cleaned periodically to ensure proper functioning.

4-8 REVIEW QUESTIONS

1. What are the three first step requirements in the installation of a lawn sprinkler system?
2. List the major types of sprinkler heads.
3. Which sprinkler heads are usually placed in corners of all perimeter areas?
4. A 1-in. (25.4 mm) supply line and a 50-psi water pressure for 28 sprinklers requires how many valves?
5. What is a sprinkler group?
6. Where is it best to start when laying out a piping system?

5 Plumber's Tools

5-1 BASIN WRENCH

The basin wrench (Fig. 5-1) is used primarily for the removal of nuts and small piping pieces from the underside of sinks and lavatories or for the removal of these same units from inaccessible spaces.

The basin wrench consists of a solid metal shaft with a sliding pin handle at its heel. At the head of the shaft is an adjustable semicircular grooved vise that may be rotated either to tighten or loosen nuts and small pipe.

5-2 OPEN END WRENCHES

Open end wrenches, such as shown in Fig. 5-2, should be part of the plumber's tool chest. These wrenches are most useful when removing small lock nuts, bolts, etc., from areas that are easily accessible. A good set of wrenches, in sizes from ¼ to ¾ in. are sufficient for most plumbing work.

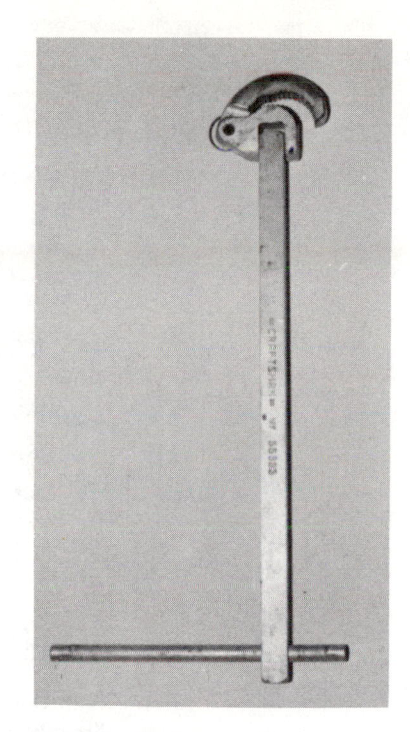

Fig. 5-1 Basin wrench.

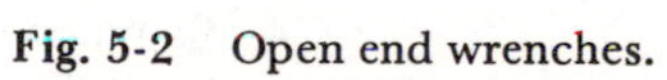

Fig. 5-2 Open end wrenches.

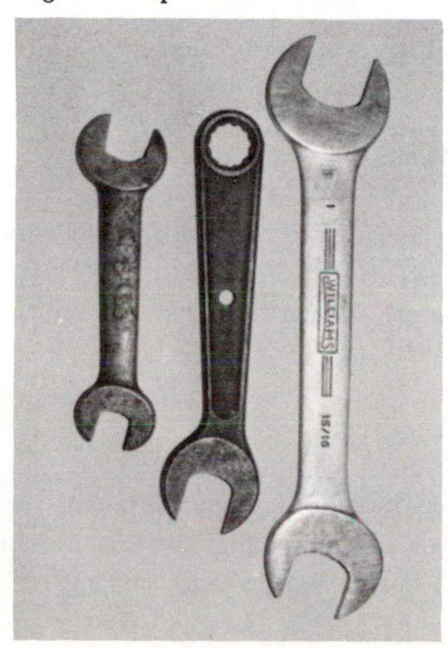

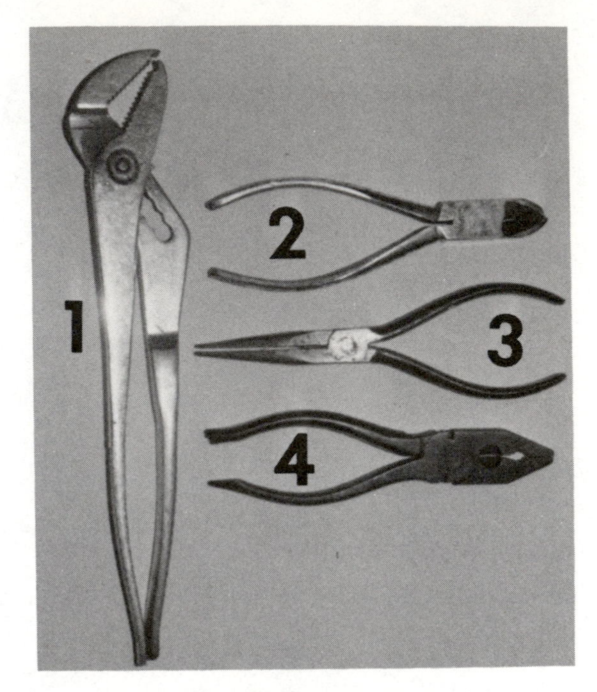

1. Channel lock pliers
2. Wire cutters
3. Long nose pliers
4. Common pliers

Fig. 5-3 Pliers.

5-3 COMMON PLIERS

Pliers (Fig. 5-3) are useful for removing and installing small nuts, fittings, etc. When working in confined areas, pliers are indispensable, as the use of conventional wrenches is not practical.

5-4 TIN SNIPS

For cutting light sheet metal up to 1/16 in. (1.6 mm) thick, the hand snip or tin snip (Fig. 5-4) is indispensable. The straight hand snip has straight blades and cutting edges sharpened to an 85° angle.

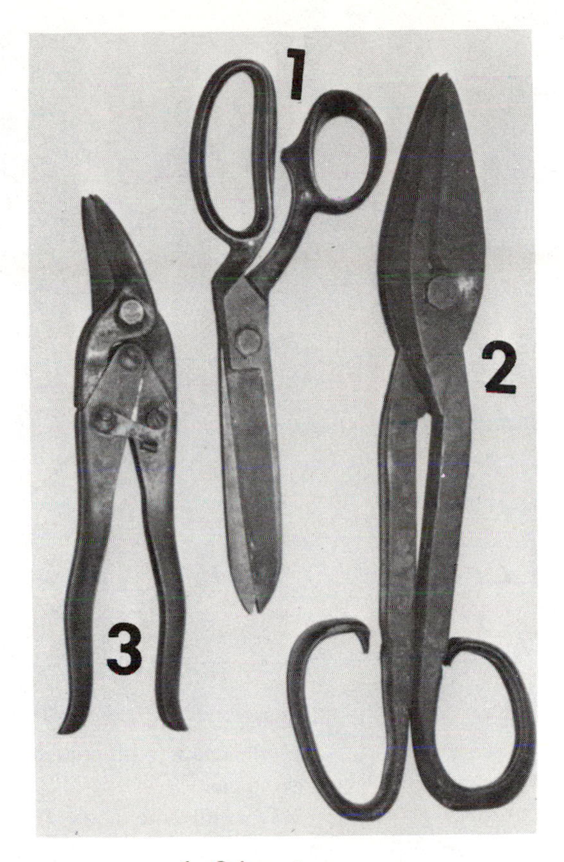

1. Scissor type
2. Straight hard snips
3. Circle snips

Fig. 5-4 Tin snips.

This type of snip is available in various sizes ranging from 6 to 14 in. (152 to 356 mm) in length.

When using tin snips it is good practice to cut somewhat outside the layout line allowing the opportunity to dress the cutting edge while keeping the material within required dimensions.

Many snips have small serrations on the blades preventing backslipping while cutting. There are special types of hand snips used for special jobs. For example, the circle snip is used for cutting out circles and elliptical shapes.

1. Thumb screw (with adjustable jaw opening)
2. Spring jaw
3. Pipe or Stillson wrench

Fig. 5-5 Pipe adjustable open end wrenches.

5-5 PIPE AND ADJUSTABLE OPEN END WRENCHES

The pipe wrench (Fig. 5-5) is primarily used for connecting and disconnecting threaded pipe and fittings. Convenient plumbing work sizes of the pipe wrench range from 6 in. to 2 ft (152 to 610 mm).

When using the pipe wrenches to install or remove pipes, they should always be used in pairs of the same size or nearly the same size.

Adjustable open end or crescent wrenches are available in a wide variety of sizes. The most common sizes range from 8 in. to 1 ft (203 to 305 mm) in length.

Adjustable open end wrenches differ from pipe wrenches in

that their jaws have a smooth finish, as opposed to the grooved jaws of pipe wrenches. They are mostly used on stainless steel or chrome finished nuts. The smooth jaws will not mar the finish of the material.

5-6 HAND DRILLS

The hand brace and the hand drill (Fig. 5-6) are essential tools of the plumber. There are two kinds of drills, the drill for making holes in metal and the drill for boring wood.

Drills for metal drilling come in numerous fractional sizes, number sizes ranging from 1 (0.228 in.) to 80 (0.0135 in.), and letter sizes ranging from A (0.234 in.) to Z (0.413 in.) The dimensions are the diameter of the hole produced. Metric-sized drills are also available.

Fig. 5-6 Hand drills.

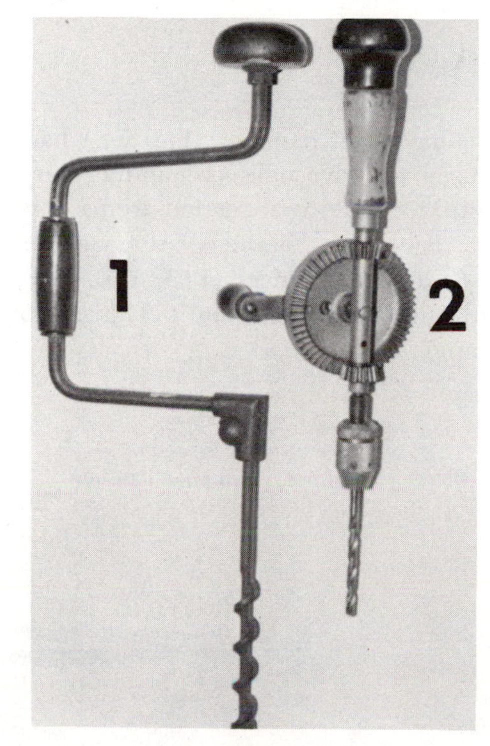

1. Hand brace
2. Hand drill

Fig. 5-7 Spud wrench.

5-7 SPUD WRENCH

The spud wrench (Fig. 5-7) is used to tighten or loosen the lock nut under a sink or lavatory. The lock nut screws onto the strainer body of the sink or lavatory.

5-8 BALL PEEN HAMMER

The ball peen hammer (Fig. 5-8) has a dual purpose. The flat face is used to drive nails or hammer surfaces, while its head or ball is useful for striking areas that are too small for the face to enter.

Ball peen hammers are made in different weights, usually 4, 6, 8, and 12 ounces (113, 170, 227, and 340 grams) and 1, 1½, and 2 lb (0.45, 0.68, and 0.91 kg). For most work a 1½ lb (0.68 kg) hammer is ideal.

Fig. 5-8 Ball peen hammer.

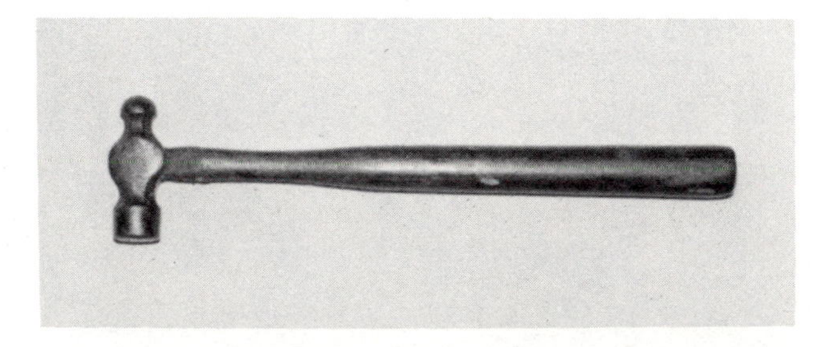

5-9 HACKSAWS

Hacksaws (Fig. 5-9) are primarily used to cut sheet metal that is too heavy for snips. Metal bar stock, as well as many types of pipe are cut with hacksaws.

Hacksaws have either an adjustable or rigid frame and a blade. Adjustable frames can hold blades that are from 8 to 10 in. (203 to 254 mm) long, while rigid saws take only one size.

The blades are made of high-grade tool steel and are tempered and hardened. The blades come in different number of teeth per inch. Use a blade of 14 teeth per inch when cutting large sections of mill material. For large sections of tough steel use a blade with 18 teeth per inch. Use 24 teeth per inch for angle iron, heavy pipe, brass, and copper, and 32 teeth per inch for thin tubing.

Fig. 5-9 Hacksaws.

14 TEETH PER INCH
FOR LARGE SECTIONS

18 TEETH PER INCH
FOR LARGE SECTIONS OF TOUGH MATERIALS

24 TEETH PER INCH
FOR ANGLE IRON, HEAVY PIPE, BRASS, COPPER

32 TEETH PER INCH
FOR THIN TUBING

TYPES OF HACKSAW TEETH

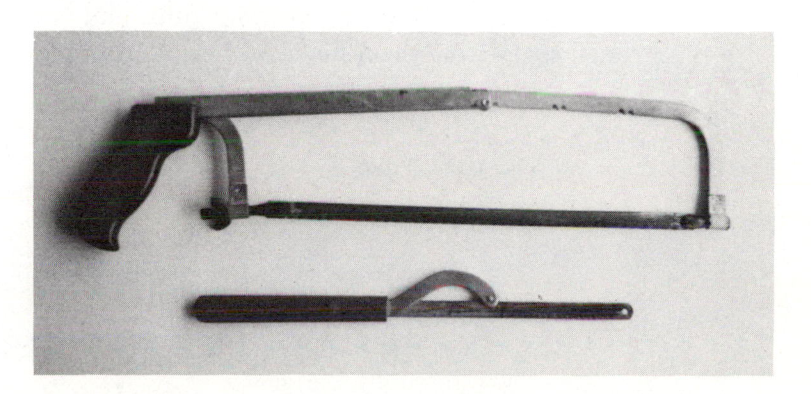

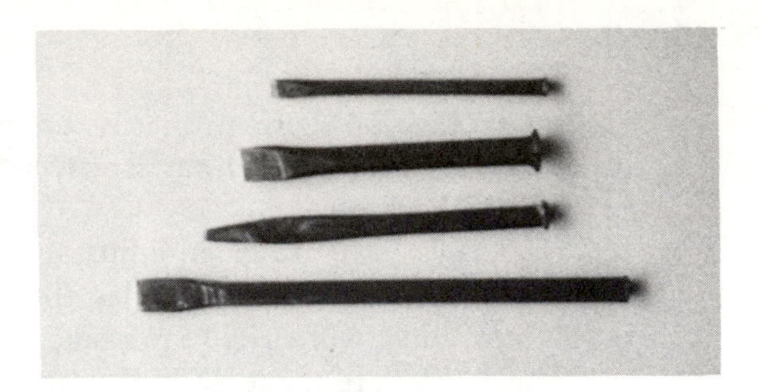

Fig. 5-10 Chisels.

5-10 CHISELS

Chisels (Fig. 5-10) are primarily used for chipping or cutting metal and wood. They are made from a good grade of tool steel. Cold chisels are classified according to the shape of their points; the width of the cutting edge indicates their size. The common shapes of chisels are flat (cold chisel), round nose, diamond point, and cape.

The flat cold chisel cuts rivets, splits nuts, chips castings, and cuts cast iron pipe and thin metal sheets. The cape chisel is used for cutting keyways or narrow grooves. Round nose chisels cut circular grooves while the diamond point chisel is used for cutting V-grooves.

5-11 FILES

Files (Fig. 5-11) are made with single-cut or double-cut teeth. Single-cut files have rows of teeth cut parallel to each other. Such files are used for finish filing, sharpening tools, and are good for taking off rough edges from sheet metal and burrs from pipe.

Fig. 5-11 Files.

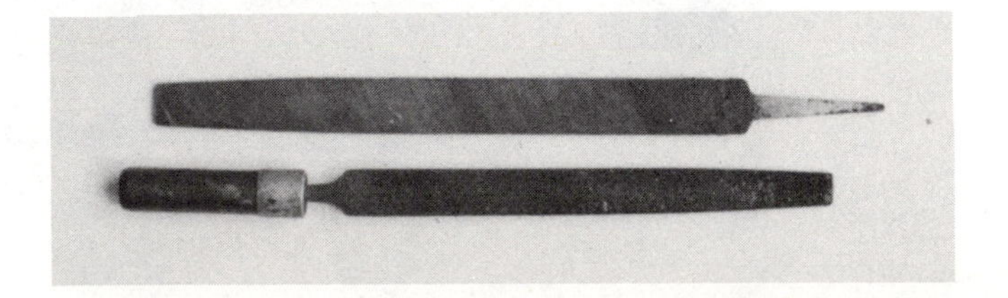

Double-cut files have crisscrossed rows of teeth forming teeth that are diamond shaped. They are used for rough work and fast cutting.

Files are graded according to the fineness or roughness of their teeth. This is largely influenced by the length of the file which typically ranges from 3 to 18 in. (76 to 459 mm).

5-12 SCREWDRIVER

The screwdriver (Fig. 5-12) is perhaps the most incorrectly used tool, often being used as a chisel or punch or scraper. However, its only true purpose is to drive and remove screws. When using the screwdriver, the most important thing is to select the proper size so that the blade fits the screw slot. There are two major types of screw drivers, the flat shank and the Phillips.

The Phillips screw has a cross-shaped slot into which the Phillips screw driver fits. There are many sizes of screwdrivers, the larger ones for large size screws, and smaller ones for smaller screws. Some screwdrivers are very short and can be used in very narrow spaces.

Fig. 5-12 Screwdrivers.

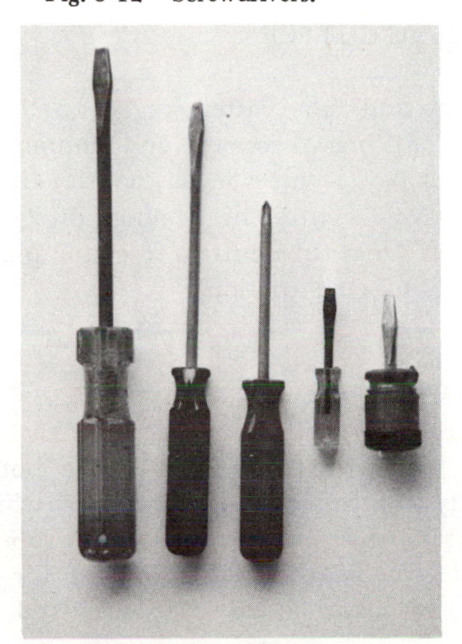

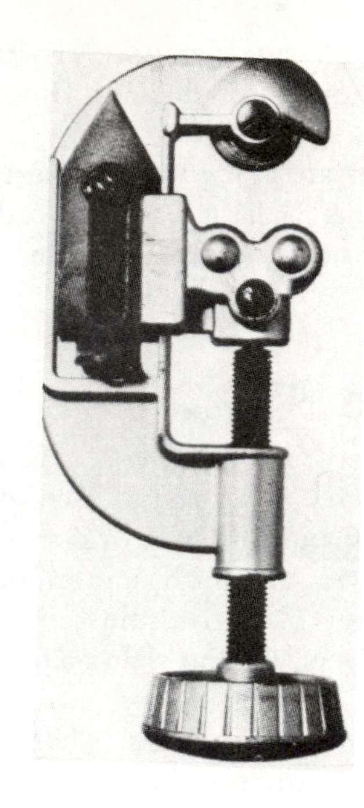

Fig. 5-13 Pipe and tubing cutter.

5-13 PIPE AND TUBING CUTTER

The pipe and tube cutter (Fig. 5-13) is made to cut pipe made of steel, iron, brass, copper, and aluminum. The hand pipe cutter known as No. 1 cutter has a capacity of ⅛ to 2 in. (3.2 to 51 mm), while a No. 2 cutter has a range of 2 to 4 in. (51 to 102 mm) in diameter. Most tube cutters resemble pipe cutters, except that they are of lighter construction.

5-14 PROPANE TORCH

The propane torch (Fig. 5-14) is a bottled gas unit containing a specially controlled air-fuel mixture which produces a high speed swirling flame. It is used for sweating copper or brass tubing joints and for soldering work. It is also used for repairing gutters, thawing pipes, and has many other uses around the home, shop, and farm.

Fig. 5-14 Propane torch.

5-15 VISE

The vise is used for holding down pipe when it is sawed, drilled, or threaded. The vise (Fig. 5-15) consists of a fixed lower jaw (5) upon which the pipe rests. The upper jaw (7) is brought down by turning the handle (1). By raising the locking lever (3) the work can quickly be removed.

5-16 PLUMBER'S MEASURING RULES

There are a number of measuring rules that are essential to the plumber. Some of the more important ones are those shown in Fig. 5-16. These are the 6-ft zig-zag rule, the flexible push-pull rule, the steel or fiber tape, and the straight steel rule.

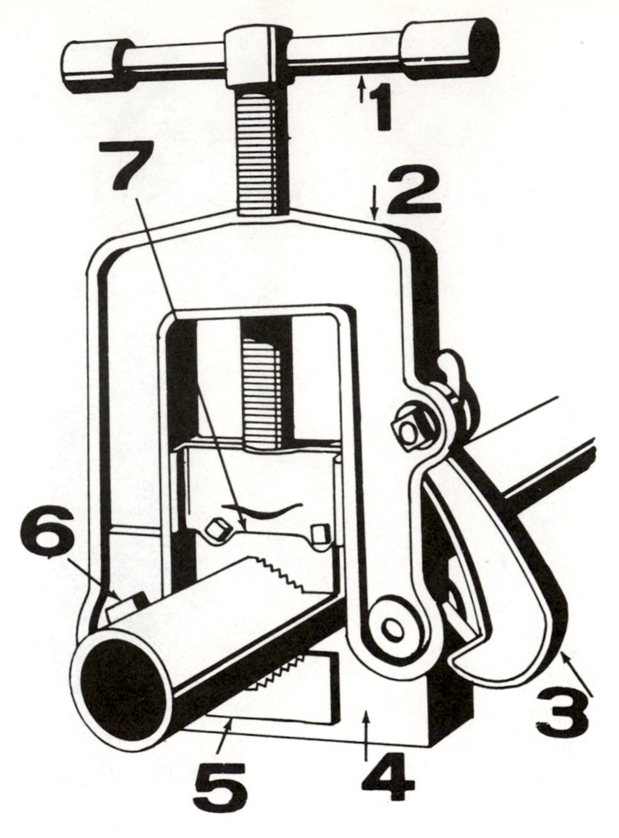

1. Handle
2. Upper body
3. Locking lever
4. Lower body
5. Lower jaw
6. Pivot
7. Upper jaw

Fig. 5-15 Vise.

5-17 OTHER USEFUL TOOLS

Level

This useful tool (Fig. 5-17) is used to level a plane in either a true horizontal or vertical position. It consists of a wood, aluminum, or steel frame into which a partially filled glass vial containing

Zig-zag rule

Push-pull rule

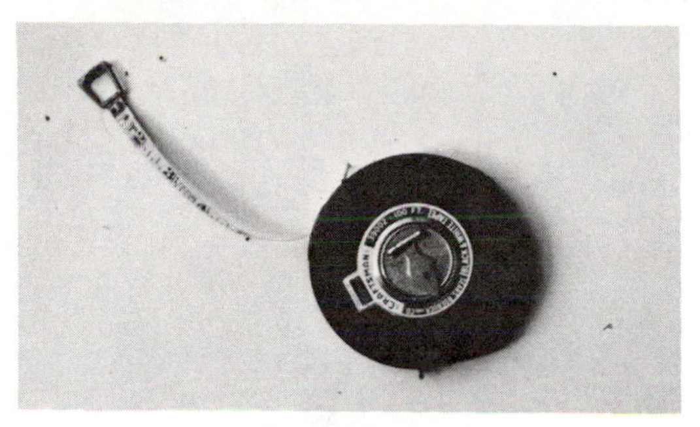

Flexible rule

Fig. 5-16 Plumber's measuring rules.

alcohol is fitted. Leveling is accomplished when the air bubble is centered between the lines.

Plumb Bob

The plumb bob (Fig. 5-17) is used to determine true verticality. The plumb bob is pointed and tapered. It is made of a brass or bronze and is suspended from a cord. Common weights are from 6 to 24 oz in increments of 2 oz.

Power Drill

The power drill (Fig. 5-18) is another useful tool for the plumber. The work to be drilled is usually held in a vise, if possible, or if the

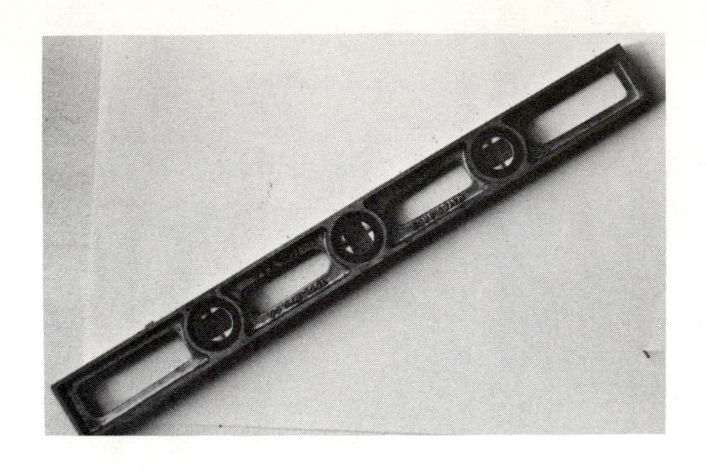

Plumb bob

Fig. 5-17 Other useful tools.

Fig. 5-18 Power drill.

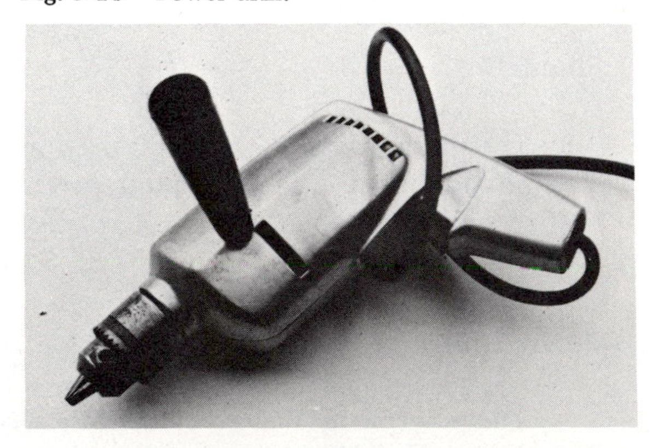

object is large enough, held tightly so it will not spin around. The exact location of the hole to be drilled must first be center-punched. The punch mark forms a starting point or seat for the drillpoint, ensuring accuracy.

Power Saw

The power saw (Fig. 5-19) with a circular blade is another useful and often used tool of the plumber. It is used for cutting wooden members through floors, walls, and partitions for pipe spaces.

Fig. 5-19 Power saw.

5-18 REVIEW QUESTIONS

1. Where is the basin wrench used?
2. What type of wrench should be used when removing nuts in easily accessible areas?
3. What is indispensible in cutting light sheet metal?
4. What are the most common wrenches used in connecting and disconnecting threaded pipe?
5. Where is the spud wrench used?
6. What must be used in cutting sheet metal that is too heavy (thick) for snips?

7. What does the flat cold chisel cut?
8. By what name are the cross-shaped slotted screws known?
9. Name three plumber's measuring rules.
10. What is the purpose of the plumb bob?

6 Plumbing Repairs

6-1 WATER OR FLUSH TANK

The homeowner can often correct most of the problems that occur within the flush tank, providing he or she clearly understands the function or purpose of the various parts within the tank. Faulty or worn parts can be removed and easily replaced by purchasing new parts from a plumbing supply house. This is many times less costly than calling a plumber. Following are some of the common breakdowns that occur and how to go about correcting them. Refer to Fig. 6-1 for identification of parts.

6-2 FLUSH LEVER

If the problem is that the tank does not flush when the flush tank lever is pressed, remove the tank cover and raise the ball stopper to let the water out. Lift the flush lever up and down and see if the lever rod also moves at the same time. If it does not, check the connection between the handle and the lever arm which may have loosened or disconnected. In some of the more modern tanks, the lever handle and lever rod are made of plastic and they may have broken. In this case purchase and install a new lever handle and

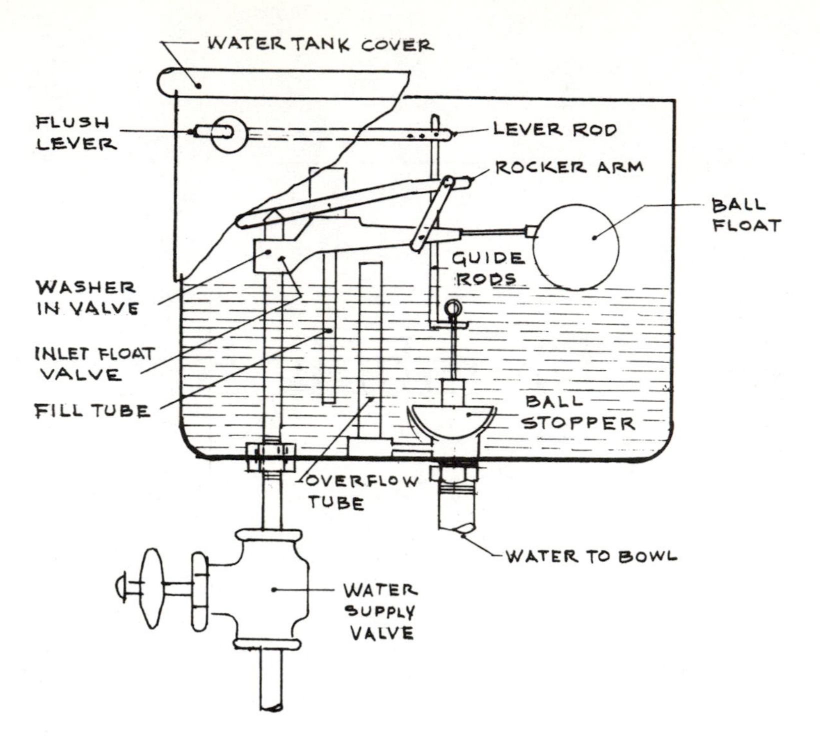

Fig. 6-1 Water tank.

lever rod of the same kind and size. Take the old parts to your plumbing supply house for comparison.

6-3 GUIDE RODS AND BALL STOPPER

The guide rods should first be checked to see if they are connected to each other and to the lever rod. There is an upper and a lower guide rod. Flush the tank and turn off the incoming cold water valve under the tank. Examine the guide rods carefully, also checking the rubber ball stopper assembly. Make certain that the upper guide rod is secured to the lever rod and that the lower guide rod is secured to the ball stopper.

The rubber ball stopper should be firm but not too soft in spots. It should not be rough or show cuts, wear marks, or have a hole in it. If the rods are found to be disconnected, simply connect them and the problem is solved. If the lower rod does not connect properly to the ball stopper, check to see if the threads on the rod and ball stopper are worn. In this case replace both rod and stopper.

6-4 OVERFLOW TUBE

The overflow tube can develop leaks and cause water to run continuously. To correct this problem, flush the tank and turn off the incoming cold water valve under the tank. Disengage the upper guide rod from the lever rod. Care must be taken not to damage the guide rod during its removal.

Lift out the entire guide rod and rubber ball stopper assembly. Do not try to shortcut this work, for trying to remove the overflow tube with the guide rods in place will cause distortion to the rods and a new problem will have been created. Next, remove the overflow tube by hand turning it counter clockwise while looking into the tank.

Get a new overflow tube of the same size from your supply store and install it in the same manner. Replace the ball stopper and guide rod assembly. Attach the upper guide rod to the lever rod. Turn on the water valve, allow the tank to fill, flush, and test for leaks.

6-5 INLET FLOAT VALVE

When the flush tank inlet float valve is not functioning, the water will be running continuously causing overfilling of the tank. To correct this problem, proceed by draining the water from the tank. Then remove the bolts from the rocker arm and from the connecting piece screwed to the float arm. Take out the stopper plug and washer from the inlet float valve. Remove the fill tube by unscrewing it by hand.

Thoroughly clean the fill tube with a thin wire rod and replace it again by hand, turning it onto the float valve. Place a new washer at the top of the inlet float valve, and then replace the brass stopper plug. Replace the rocker arm and the connecting piece that fastens it to the float arm. Allow water into the tank, flush, and check if fault has been corrected.

6-6 BALL FLOAT

In order to check the ball float, flush the tank and shut off the incoming water valve. Carefully unscrew the ball float from the rod and inspect it for holes, cracks, or stripped threads. If there is any damage, replace it with a new ball float. Floats are available in copper or plastic. Secure the new float by holding the rod with one hand and screwing the ball float on with the other. Care should be taken not to bend the rod, or force the threaded parts. Hand tightening the rod to the ball float is all that is necessary.

Reopen the water valve and fill the tank. Check to see that the ball float stops the inflow of water to the tank before the water level reaches the top of the overflow tube.

6-7 WATER SUPPLY LINE INLET

Leaks at the water supply line coming into the flush tank at the bottom of the tank are caused by worn out washers. Sometimes a slight tightening of the nut will stop the leak but this can be dangerous if the nut is over tightened. If the leak persists, empty the tank of water, and turn off the incoming water supply valve. Loosen the nuts both inside and outside the tank and replace the old washer with a new one. Turn on the nuts and tighten them, being careful not to over tighten them. Turn on the valve and allow the tank to fill. Check to see if the leaks have stopped.

6-8 REPAIRING A FAUCET LEAK

When installing a faucet onto a lavatory or sink (Fig. 6-2), apply plumber's putty all around the bottom groove of the faucet and place the faucet shank through the hole in the lavatory or sink. Slide on the washer and then tighten the lock nut. Next, remove

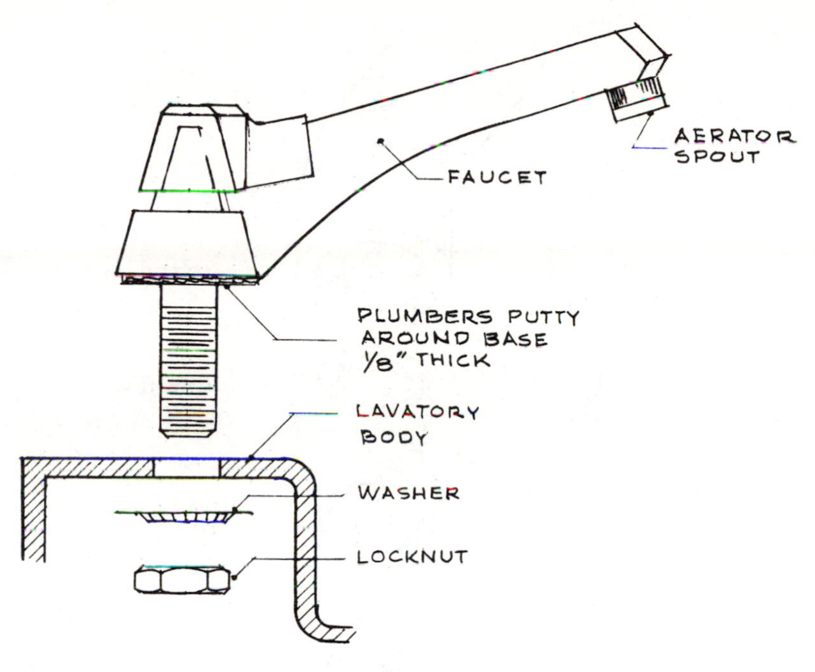

Fig. 6-2 Fixing faucet leak at base.

excess putty from the base of faucet. Finally connect water supply to shank.

6-9 AERATORS

The aerator (Fig. 6-3) can be unscrewed from the end of the spout for periodic cleaning. Small particles of grit in the water supply will clog the aerator. The aerator consists of a series of small parts, which, when taken apart should be set aside in correct order so that when reassembling all parts are correctly assembled.

The aerator shown here consists of a plastic or rubber washer, a disk perforated by tiny holes, and a screen. Clean the screen with a small stiff brush; also use the brush and a tooth pick to clean the disk and intake holes. Replace badly worn washers, and flush out all parts by holding them upside down in a full stream of water before reassembling. Follow the exploded drawing of the six-part aerator and its assembly.

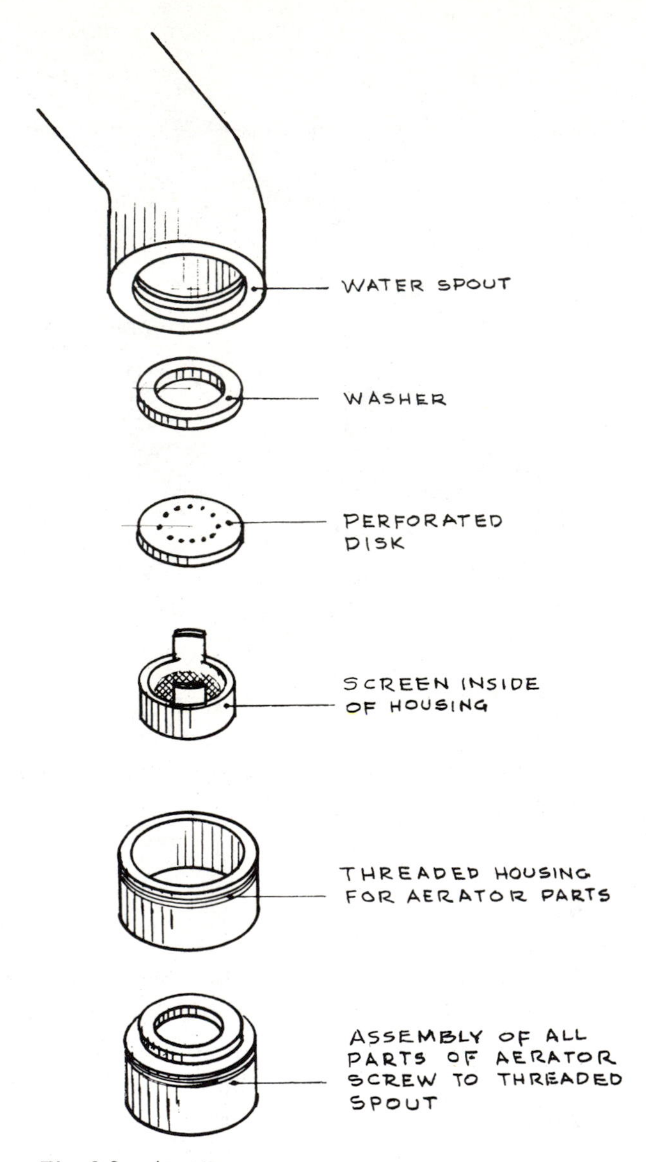

Fig. 6-3 Aerator.

6-10 VALVE SEAT

Faucet leaks are often caused by worn valve seats (Fig. 6-4). Many valve seats are threaded into the body of the faucet and can be easily replaced. First, remove the handle, then the bonnet cap,

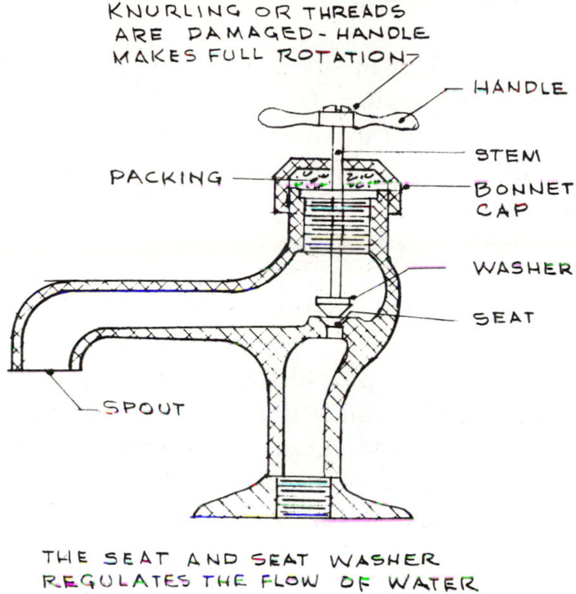

Fig. 6-4 Faucet showing valve seat.

and lift out the stem, at the end of which is the washer and the washer retainer. If the washer is worn, replace it with a new one.

Use a flashlight to inspect the seat ring for rough edges, burrs, or other distortions. If the seat ring is worn, remove it with a seat wrench providing of course, the seat ring is threaded into the faucet body. Take it to your plumbing supply store to get an exact duplicate. Lubricate the outside of the new seat with pipe joint compound and push it firmly onto the seat wrench. Screw it into the faucet body. If the valve seat cannot be removed, its surface must be ground smooth with a valve-seat dresser.

Stripped Faucet Handle

Sometimes the handle knurling is stripped and the faucet cannot be entirely shut off. A new handle may normally be purchased at any plumbing supply store or the problem can be fixed temporarily by placing tape around the stem.

Bonnet Packing

Leaks can often occur around the bonnet cap indicating worn bonnet packing. To remedy this situation, remove the handle and the bonnet cap and replace the bonnet packing with new packing. Sometimes a slight tightening of the bonnet cap may stop the leak.

6-11 LAVATORY POP-UP DRAIN

The pop-up plug drain (Fig. 6-5) can be closed by lifting the lift rod. The lift rod is attached to the clevis rod which in turn is attached to a pivot rod. Leaks often occur around the plastic pivot ball and they can easily be corrected by tightening the nut of the pivot ball housing. If the leak does not stop, replace the pivot ball washer.

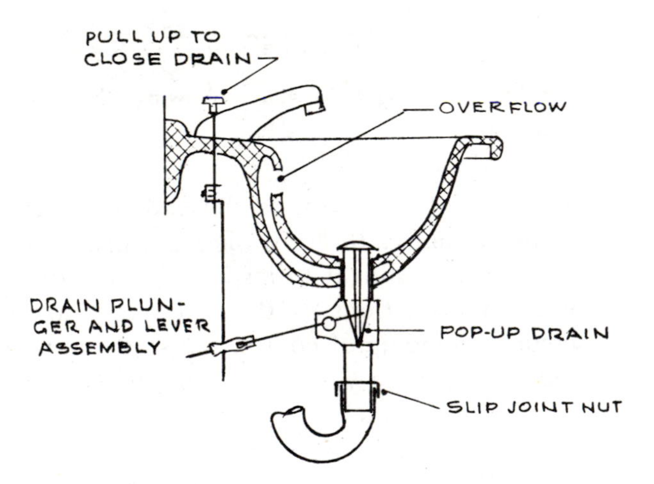

Fig. 6-5 Lavatory pop-up drain.

Another malfunction often occurs when the lift rod does not close the pop-up plug. In this case check the clevis rod and pivot rod assembly. You may find that a small screw has loosened and the rod lifts out of the assembly. Tighten this screw when the pull-up is fully depressed and the pop-up plug is open or raised.

6-12 LAVATORY VALVE INSTALLATION

To install a shutoff valve on both hot and cold water lines under an existing lavatory (Fig. 6-6) turn off the house water line and open the lavatory faucet to let out excess water. With a hacksaw cut a

Fig. 6-6 Lavatory valve installation.

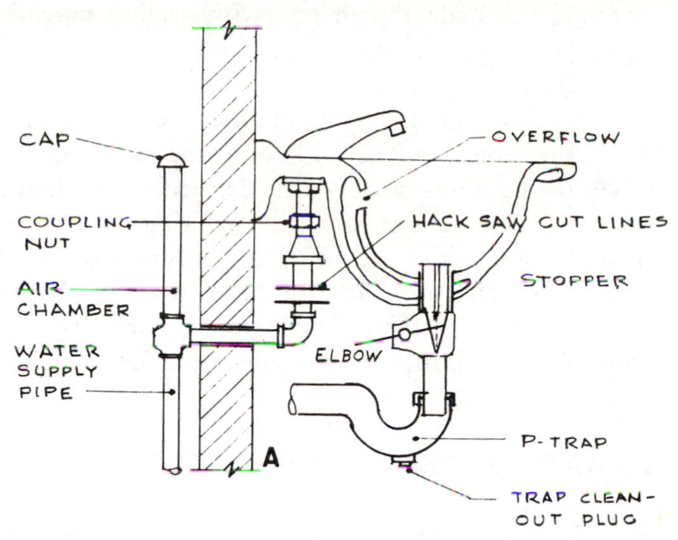

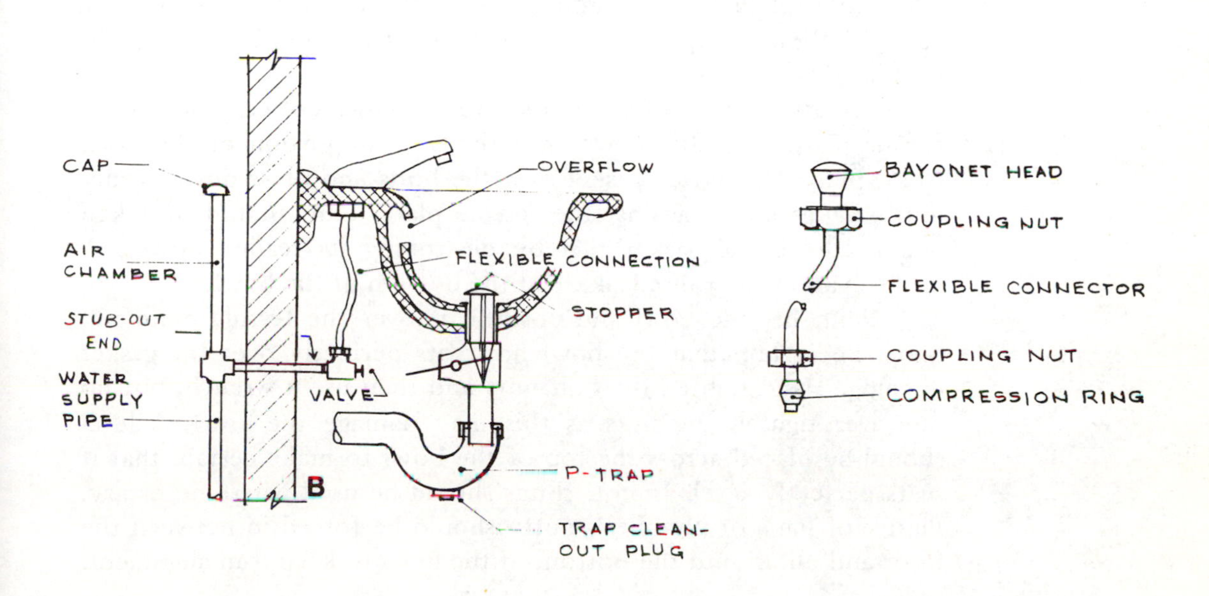

½-in. piece, as shown at A, out of the supply pipe indicated by lines through the pipe. With a basin wrench loosen the coupling nut under the lavatory faucet and let it slide down on the tail piece which at the same time will cause the tail piece to drop out.

Use two pipe wrenches to unscrew the elbow. Hold one wrench stationary on the stub-out pipe while turning the elbow with the other. Screw the new stop valve on the threaded stub-out end after wrapping the male thread end with sealing tape. Tighten the valve connection with a wrench. The outlet of the valve should be directly below the fixture inlet.

Use the proper length flexible connector, making its head fit the fixture inlet and its end fit inside the stop valve outlet, as at B. Slip the coupling nut over the connector and join the head and the inlet. Slide the compression nut and ring over the connector's plain end and insert it into the outlet and secure it by hand. Finish tightening the coupling nut with the basin wrench, then tighten the compression fitting with the adjustable wrench. To test for leaks turn the water on and inspect all connections for leaks.

6-13 SETTING A WATER CLOSET TO THE FLOOR

When installing a new water closet (toilet bowl) (Fig. 6-7), check the ferrule plate to be certain that it is clean and level. Also check to see if the entire perimeter of the plate is securely fastened to the lead bend.

Slip the threads of the bolts into the large slotted holes in the ferrule plate, and slide them into the narrow portion of the slots. Next, place the ferrule gasket over the bolts and press down firmly until the gasket is against the ferrule plate. Care should be taken not to break the gasket thereby destroying its sealing power. A broken gasket can cause leakage at the bottom of the bowl.

Next, lift the bowl and position it over the ferrule plate and bolts. Be certain that the bowl horn fits perfectly over the gasket opening. Hand tighten the bolt nuts and then use a wrench, but do not over tighten the nuts as this may damage the bowl. A level should be placed across the top of the bowl to make certain that it seats perfectly level. If not, shims should be used where necessary. Plaster of Paris or plumber's putty should be forced in between the floor and all around the bottom of the bowl to keep it in alignment.

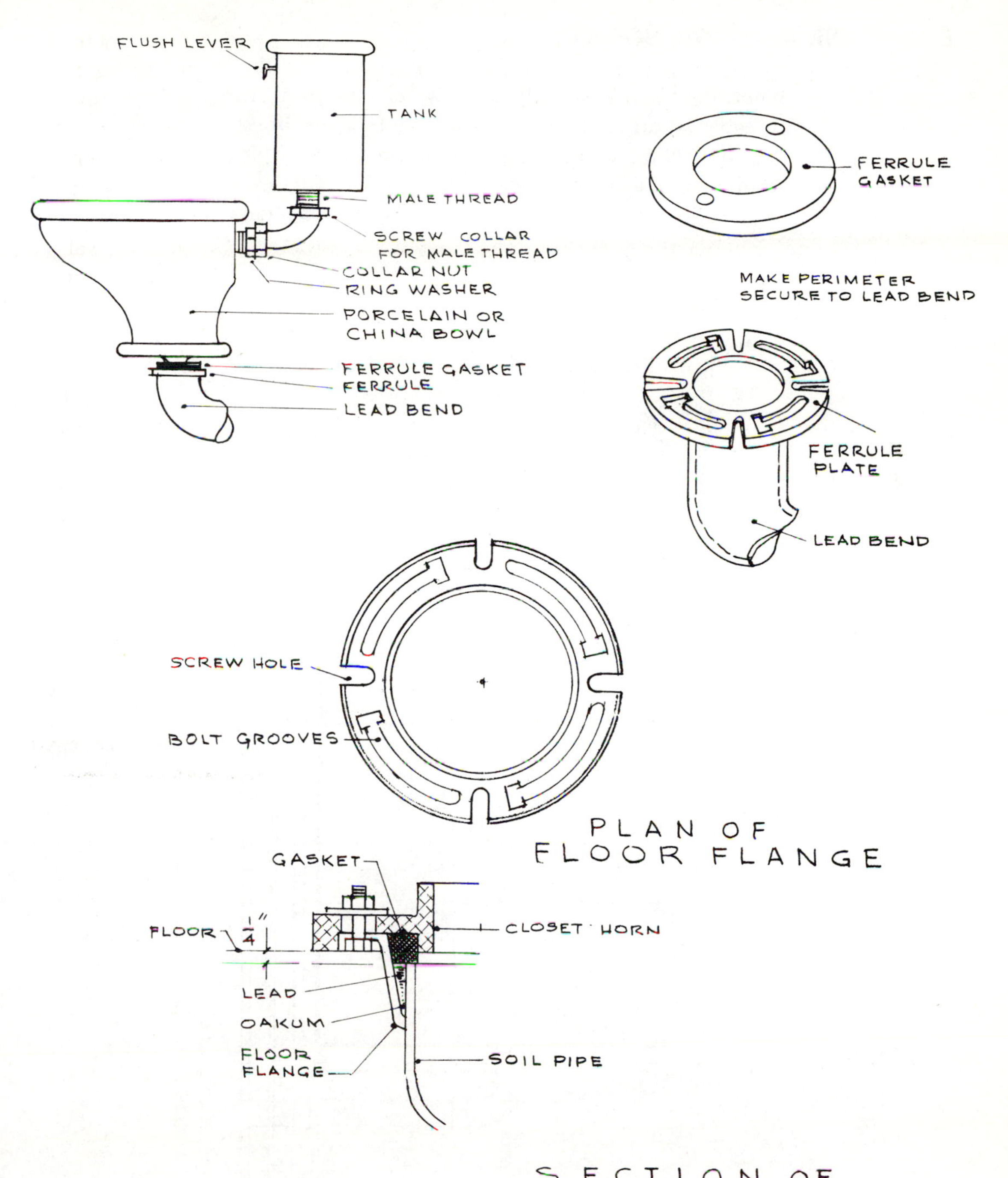

Fig. 6-7 Setting water closet.

6-14 TANK AND BOWL HOOK UP

When the bowl is set, place the water tank in position and secure the threaded spud, the spud washer, and flange brass nut, as shown in Fig. 6-8. Place compound or tape on the male end of the threaded spud, and secure the slip ell with its collar nut.

To complete the installation, secure the tank to the wall and

Fig. 6-8 Tank and bowl hook up water connection.

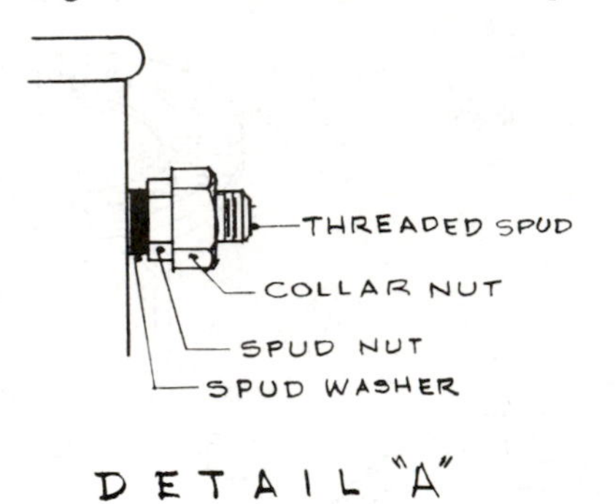

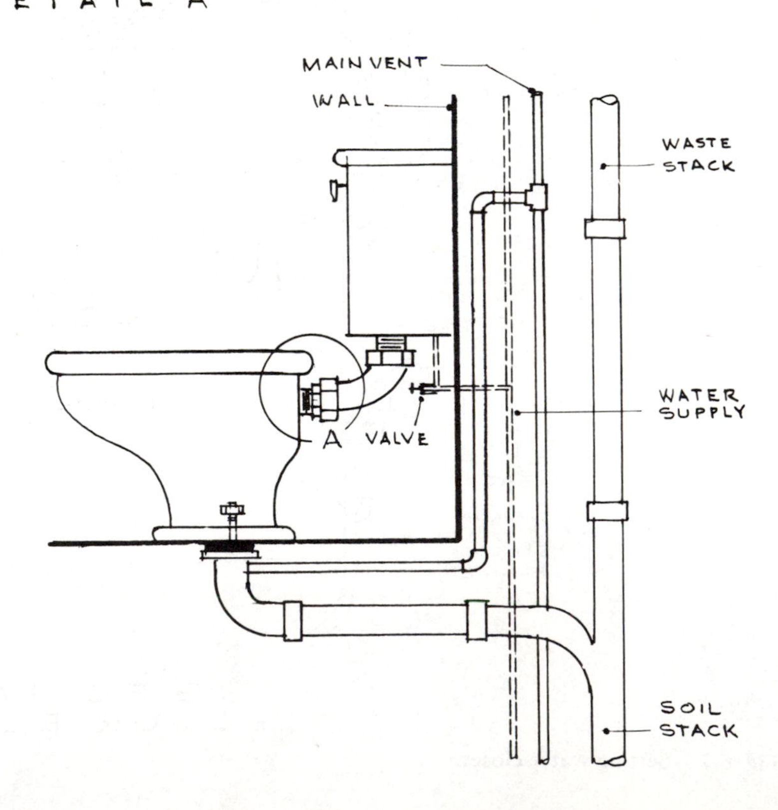

then connect the upper end of the slip ell to the tank male threaded outlet at the bottom. Use compound or tape on the male threads only.

Figure 6-8 also shows the cold water supply line to the flush tank, the vent hook-up for the water closet, and the water closet branch line into the soil stack.

6-15 INSTALLING A TANK MOUNTED ON A BOWL

The first step is to fit the beveled rubber spud washer into the water inlet hole of the water closet. See Fig. 6-9. Next, place the tank cushion over the rear part of the bowl and let the holes in the cushion

Fig. 6-9 Tank mounted on bowl.

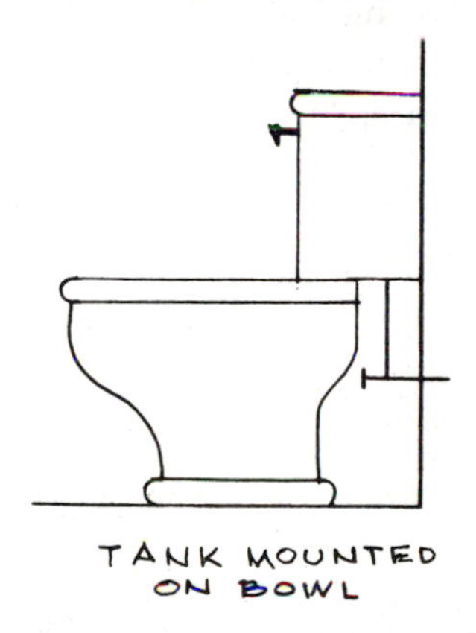

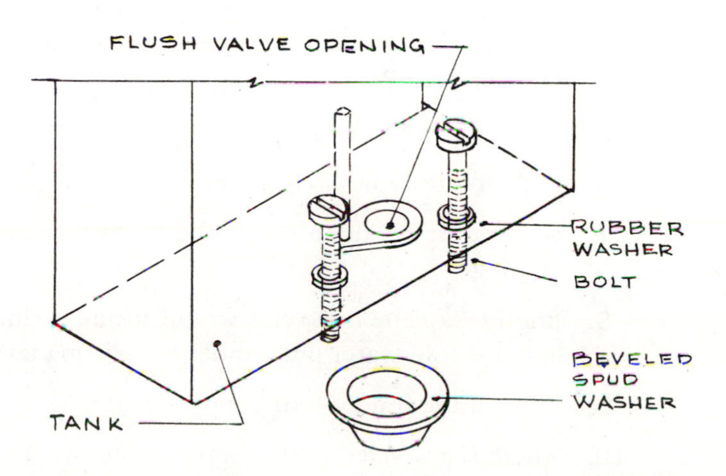

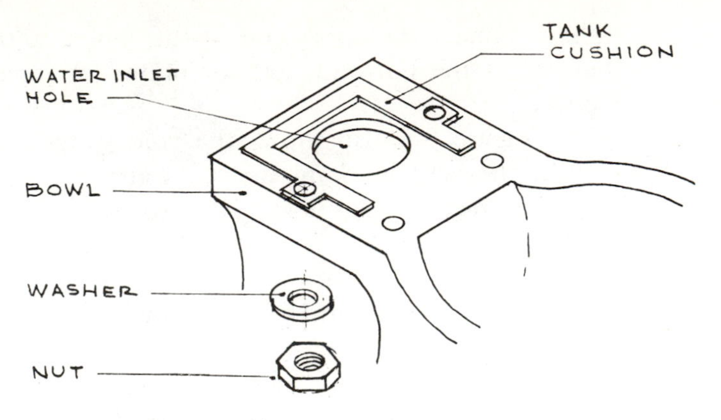

Fig. 6.9 (Continued)

align with those of the bowl and tank. The tank is then placed on the tank cushion allowing the bolts to pass through the cushion and the sides of the toilet bowl. The bolts, equipped with rubber washers inside the tank, fasten the tank to the bowl.

6-16 REVIEW QUESTIONS

1. If the tank does not flush when depressing the lever handle, what steps would you take to find the possible causes?
2. What steps must be taken when water in the tank runs continuously?
3. Explain the procedure in checking the ball float for leaks.
4. A leak has developed at the bottom of the incoming supply line under the flush tank. What causes this leak and how would you correct it?
5. List the parts of a faucet aerator.
6. Explain how a new valve seat is installed.
7. What must be done when the lift rod does not close the pop-up plug?
8. Briefly explain how you would install a shutoff valve on both hot and cold water lines under an existing lavatory.
9. Explain the function of a ferrule plate.
10. Where is a beveled rubber spud washer used?

7 Heat Loss: How To Compute It

7-1 WHAT IS HEAT LOSS?

Heat loss represents the quantity of heat that the heating system must supply to a given space to maintain the required temperature within that space.

7-2 WHY DOES A ROOM HAVE A LOSS OF HEAT?

It is a well-known fact that heat flows "downhill," that is, from a high-temperature source to a low-temperature receiver. Thus warm objects lose their heat to colder objects. In the case of a heated structure, the room air acts as the high-temperature source and the outside air as the low-temperature receiver, with the building wall serving as a barrier between the two. (*A loss of heat will take place as long as a temperature differential exists.*) It seems logical to conclude that the amount of heat lost will depend primarily on the following considerations:

1. How much surface is exposed to a cold temperature?
2. How good a heat barrier is this surface?
3. How great is the temperature difference causing the heat flow?

It follows directly then that:

1. The larger the room, the greater the heat loss.
2. The poorer the wall construction, the greater the heat loss.
3. The colder the outdoor temperature, the greater the heat loss.

7-3 WHAT IS THE BASIC RULE FOR COMPUTING HEAT LOSS?

Heat loss = Area X Wall effectiveness X Degree temperature difference

Since heat loss depends on surface, wall construction, and temperature differential, the actual heat loss is expressed in terms of area, wall effectiveness, and degrees of temperature difference between inside and outside design temperatures. In the above expression:

The area refers to the square footage of only those surfaces that are exposed to a temperature less than room temperature. By surface we mean wall, glass, door, floor, roof, or partition area. It should be noted that no heat loss occurs between adjacent rooms if each room is heated to the same temperature. The symbol for area is A.

Wall effectiveness expresses the ability of a material to resist the flow of heat. A heat-flow transmission coefficient has been assigned to every type of commercially used surface. Constructions with low transmission coefficients indicate good insulators; those with relatively high factors indicate poor insulators. These factors are variously referred to in the trade as heat-transmission coefficients or U factors. The standard heat-transmission coefficient symbol adopted by the American Society of Heating and Ventilating Engineers is U. Table 7-1 shows a few typical heat-transmission coefficients.

Glass for example, is a poor insulator, because it has a relatively high transmission coefficient. It will be noted that the addition of rock-wool insulation between the studs of the exterior wall shown cuts down the heat loss by more than half. Heat-transmission coefficients for ordinary residential construction vary from 0.07 for well-insulated walls to 1.13 for glass and thin wood doors. Com-

TABLE 7-1

Typical Heat Transmission Coefficients

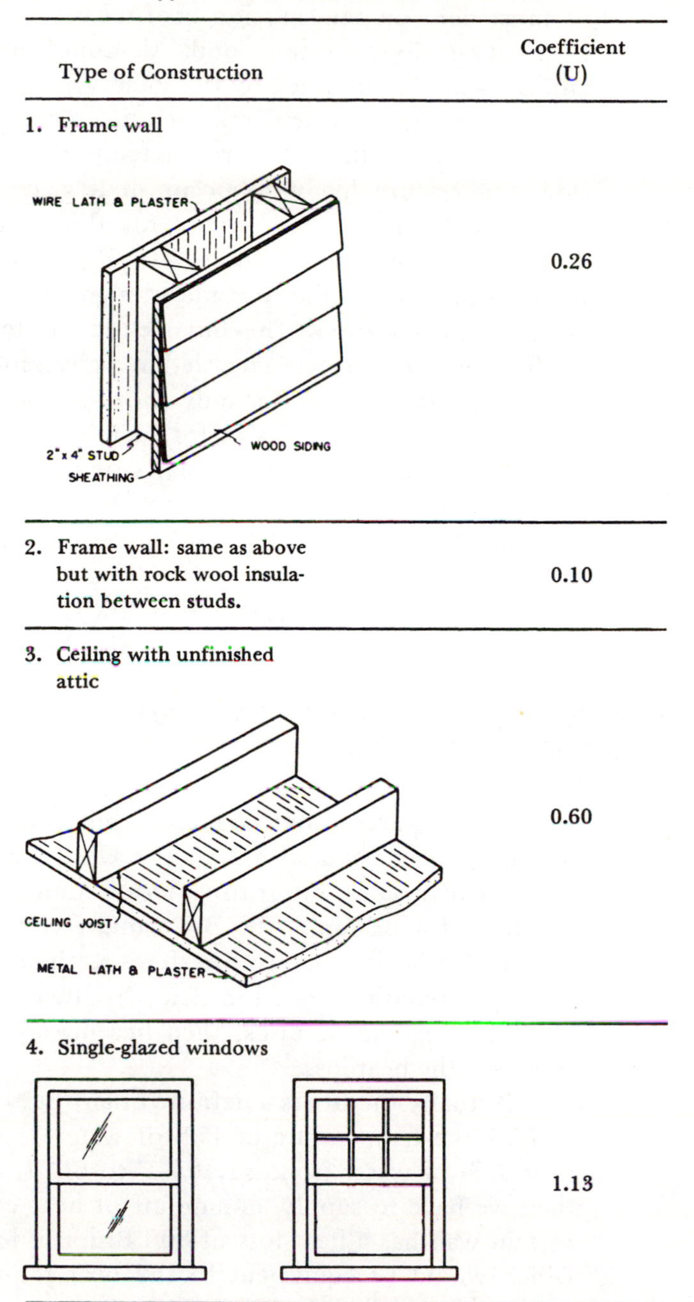

Type of Construction	Coefficient (U)
1. Frame wall	0.26
2. Frame wall: same as above but with rock wool insulation between studs.	0.10
3. Ceiling with unfinished attic	0.60
4. Single-glazed windows	1.13

plete tabulations for all building materials may be found in any standard engineering text dealing with heating. (The original source of most data on transmission coefficients is the American Society of Heating, Refrigerating, and Air-Conditioning Engineers Guide and Data Book, published by the society.)

The degrees of temperature difference is the arithmetic difference between the inside room temperature and the coldest outside temperature likely to occur in a given locality, expressed in degrees Fahrenheit. The symbol for this temperature difference is T. It is common practice to use 70° F (21° C) for the inside design temperature, with the outside design temperature varying with geographical location. The outside design temperature is not the coldest temperature ever recorded in a given locality, since extremely low temperatures exist for only short periods of time and it would not be economical to design for such a condition. Table 7-2 shows design temperatures in various parts of the country. Note how they vary considerably.

The basic rule for heat loss may now be stated:

$$\text{Heat loss} = A \times U \times T$$

7-4 WHAT UNIT OF MEASUREMENT IS USED TO EXPRESS HEAT LOSS?

Heat loss requires a unit of measurement, as do length, area, weight, distance, speed, and so on. In the United States and Great Britain the unit of heat is the British thermal unit, abbreviated Btu. Unlike the foot, for example, the Btu cannot very well be visualized, but this limitation in no way interferes with its usefulness. As long as heating apparatus is rated in Btu's by the manufacturer and the heat losses are figured in Btu's, then heating equipment can be selected to match the heat loss.

Actually the Btu is a definite quantity of heat, the heat required to raise the temperature of 1 lb of water 1°F. In other words if we heat 1 lb of water from, say 65°F to 66°F, or from 86°F to 87°F, then we have to supply an amount of heat equal to 1 Btu. Now if a certain wall has a heat loss of 800 Btu, it is losing heat in an amount which would be equivalent to the heat required to heat 800 lb of water 1°F, 400 lb 2°F, 100 lb 8°F, and so on.

TABLE 7-2

Outside Design Temperatures in Various Parts of the United States

Location	°F	°C
ALABAMA		
Birmingham	15	−9.4
Mobile	25	−3.9
Montgomery	20	−6.7
Tuscaloosa	20	−6.7
ALASKA		
Anchorage	−25	−31.7
Fairbanks	−50	−45.6
Juneau	− 5	−20.6
Nome	−30	−34.4
ARIZONA		
Flagstaff	0	−17.8
Phoenix	30	−1.1
Tucson	30	−1.1
Yuma	35	1.7
ARKANSAS		
Fort Smith	10	−12.2
Hot Springs	15	−9.4
Little Rock	15	−9.4
CALIFORNIA		
Bakersfield	30	−1.1
Fresno	30	−1.1
Los Angeles	40	4.4
Mission Viejo	35	1.7
Sacramento	30	−1.1
San Diego	40	4.4
San Francisco	35	1.7
COLORADO		
Boulder	−5	−20.6
Denver	−5	−20.6
Grand Junction	0	−17.8
Pueblo	−5	−20.6
CONNECTICUT		
Bridgeport	10	−12.2
Hartford	5	−15.0
New Haven	5	−15.0
New London	5	−15.0
DELAWARE		
Dover	10	−12.2
Wilmington	10	−12.2
DISTRICT OF COLUMBIA		
Washington	15	− 9.4
FLORIDA		
Jacksonville	40	4.4
Miami	45	7.2
Tallahassee	25	−3.9
Tampa	35	1.7
GEORGIA		
Atlanta	15	−9.4
Augusta	20	−6.7
Macon	20	−6.7
Savannah	25	−3.9
IDAHO		
Boise	5	−15.0
Lewiston	0	−17.8
Twin Falls	−5	−20.6
ILLINOIS		
Chicago	−10	−23.3
Decatur	−5	−20.6
Joliet	−5	−20.6
Springfield	−5	−20.6
INDIANA		
Evansville	5	−15.0
Indianapolis	0	−17.8
South Bend	−5	−20.6
Terre Haute	0	−17.8
IOWA		
Ames	−10	−23.3
Burlington	−5	−20.6
Des Moines	−10	−23.3
Sioux City	−10	−23.3
KANSAS		
Dodge City	0	−17.8
Salina	0	−17.8
Topeka	0	−17.8
Wichita	5	−15.0

TABLE 7-2 (Continued)

Outside Design Temperatures in Various Parts of the United States

Location	°F	°C
KENTUCKY		
Ashland	5	−15.0
Lexington	5	−15.0
Louisville	5	−15.0
LOUISIANA		
Alexandria	25	−3.9
Baton Rouge	25	−3.9
New Orleans	30	−1.1
Shreveport	20	−6.7
MAINE		
Augusta	−5	−20.6
Bangor	−10	−23.3
Portland	−5	−20.6
MARYLAND		
Baltimore	10	−12.2
Cumberland	5	−15.0
MASSACHUSETTS		
Boston	5	−15.0
Springfield	−5	−20.6
Worcester	0	−17.8
MICHIGAN		
Battle Creek	0	−17.8
Grand Rapids	0	−17.8
Lansing	−5	−20.6
Muskegon	0	-17.8
MINNESOTA		
Alexandria	−20	−28.9
Duluth	−20	−28.9
Minneapolis	−15	−26.1
St. Paul	−15	−26.1
MISSISSIPPI		
Columbus	15	−9.4
Jackson	20	−6.7
Meridian	20	−6.7
Vicksburg	20	−6.7
MISSOURI		
Columbia	0	−17.8
Kansas City	5	−15.0
St. Louis	0	−17.8
Springfield	5	−15.0
MONTANA		
Butte	−25	−31.7
Helena	−20	−28.9
Miles City	−20	−28.9
Missoula	−15	−26.1
NEBRASKA		
Grand Island	−10	−23.3
Lincoln	−5	−20.6
North Platte	−10	−23.3
Omaha	−10	−23.3
NEVADA		
Carson City	5	−15.0
Las Vegas	25	−3.9
Reno	5	−15.0
NEW HAMPSHIRE		
Concord	−10	−23.3
Manchester	−10	−23.3
NEW JERSEY		
Atlantic City	10	−12.2
Newark	10	−12.2
Paterson	5	−15.0
Trenton	10	−12.2
NEW MEXICO		
Albuquerque	10	−12.2
Santa Fe	5	−15.0
NEW YORK		
Albany	5	−15.0
Buffalo	0	−17.8
Elmira	−5	−20.6
New York	10	−12.2
Utica	−10	−23.3
NORTH CAROLINA		
Asheville	10	−12.2
Durham	15	−9.4
Raleigh	15	−9.4
Wilmington	25	−3.9
NORTH DAKOTA		
Bismarck	−25	−31.7
Fargo	−20	−28.9
Grand Forks	−25	−31.7
Jamestown	−20	−28.9

TABLE 7-2 (Continued)

Outside Design Temperatures in Various Parts of the United States

Location	°F	°C	Location	°F	°C
OHIO			TEXAS		
Akron	0	−17.8	Abilene	15	−9.4
Cincinnati	0	−17.8	Amarillo	5	−15.0
Cleveland	0	−17.8	Dallas	20	−6.7
Toledo	−5	−20.6	Houston	25	−3.9
OKLAHOMA			San Antonio	25	−3.9
Muskogee	10	−12.2			
Oklahoma City	10	−12.2	UTAH		
Tulsa	10	−12.2	Ogden	0	−17.8
			Salt Lake City	5	−15.0
OREGON					
Medford	20	−6.7	VERMONT		
Pendleton	0	−17.8	Barre	−15	−26.1
Portland	15	−9.4	Burlington	−10	−23.3
Salem	20	−6.7			
			VIRGINIA		
PENNSYLVANIA			Lynchburg	10	−12.2
Harrisburg	5	−15.0	Norfolk	20	−6.7
Philadelphia	10	−12.2	Richmond	15	−9.4
Pittsburgh	0	−17.8	Roanoke	10	−12.2
Scranton	0	−17.8			
			WASHINGTON		
RHODE ISLAND			Seattle	20	−6.7
Newport	5	−15.0	Spokane	−5	−20.6
Providence	5	−15.0	Tacoma	20	−6.7
			Walla Walla	0	−17.8
SOUTH CAROLINA					
Charleston	25	−3.9	WEST VIRGINIA		
Columbia	20	−6.7	Charleston	5	−15.0
Greenville	20	−6.7	Elkins	0	−17.8
			Martinsburg	5	−15.0
SOUTH DAKOTA			Wheeling	0	−17.8
Aberdeen	−20	−28.9			
Rapid City	−10	−23.3	WISCONSIN		
Sioux Falls	−15	−26.1	Green Bay	−15	−26.1
Watertown	−20	−28.9	La Crosse	−15	−26.1
			Madison	−10	−23.3
TENNESSEE					
Chattanooga	15	−9.4	WYOMING		
Knoxville	15	−9.4	Cheyenne	−10	−23.3
Memphis	15	−9.4	Laramie	−15	−26.1
Nashville	10	−12.2	Sheridan	−15	−26.1

Courtesy *ASHRAE Guide and Data Book.*

Time is also a factor to consider. If a wall has a heat loss of 600 Btu's, it is necessary to know whether this loss takes place in one minute, one hour, or one day. It is customary to express all heat losses in terms of Btu's per hour, written Btu/hr. The value of U in the heat-loss formula must therefore be expressed in the units of Btu per square foot of surface per hour per degree of temperature difference.

Is it possible to shorten the heat-loss formula for easy use? Yes. Look at the heat-loss formula once more:

$$\text{Heat loss} = A \times U \times T$$

For a given construction, U is always the same, and for a given locality the inside and outside design temperatures are fixed. Therefore the product of U and T is always a constant number. If an inside temperature of 70°F (21°C) is chosen and an outdoor design temperature of 0°F (–17.8°) is used, then T = 70° – 0° = 70°F (38.8°C). If the known heat-transmission coefficient U is multiplied by 70°F, a combined number will be found, which for convenience will be called the HTN (heat transmission number) factor. All that is necessary now is to multiply the area in question by the combined HTN factor to arrive at the heat loss in Btu/hr. Thus:

$$\text{Btu/hr} = A \times \text{HTN}$$

7-5 HOW CAN A SIMPLIFIED HEAT-LOSS TABLE BE MADE?

A simplified heat-loss table can be made by obtaining heat-transmission coefficients from any text on the subject, multiplying all coefficients by the desired temperature differential between inside and outside design temperature, and listing all factors thus obtained for all the desired types of construction. Table 7-3 has been prepared in this manner.

Notice that the HTN numbers shown in Table 7-3, section 1, wall 1, for example, can be obtained by multiplying the heat-loss coefficient U (0.26) by 70°F:

$$\text{HTN} = 0.26 \times 70° = 18.2$$

TABLE 7-3

Simplified Heat Transmission Numbers (HTN)

No.	(1) Walls	HTN	U
	Frame		
1	Wood siding or clapboard on sheathing, no insulation	18.2	0.26
2	Same, but insulated with rock wool between studs	7.0	0.10
3	Wood shingles, no insulation	18.3	0.26
4	Same, but insulated with rock wool between studs	7.0	0.10
5	Stucco, no insulation	22.4	0.32
6	Same, but insulated with rock wool between studs	8.4	0.12
7	Brick veneer, no insulation	19.6	0.28
8	Same, but insulated with rock wool between studs	9.1	0.13
	Masonry		
9	8-in. solid brick, plastered, no insulation	22.4	0.32
10	Same, but with 1-in. rigid insulation under plaster	11.2	0.16
11	12-in. solid brick, plastered, no insulation	17.5	0.25
12	Same, but with 1-in. rigid insulation, under plaster	9.8	0.14
13	8-in. concrete block, plastered, no insulation	17.5	0.25
14	Same, but with 1-in. rigid insulation, under plaster	10.5	0.15
15	Brick veneer on 8-in. concrete block, plastered, not insulated	16.1	0.23
16	Same, but with 1-in. rigid insulation	9.8	0.14
	Glass		
17	4-in. thick hollow glass block	34.3	0.49
	(2) Ceilings		
	Wood		
1	Joists and plaster, no floor above, no insulation	24.1	0.34
2	Joists and plaster, rough floor above, no insulation	10.5	0.15
3	Same, with rock-wool insulation	4.0	0.057

TABLE 7–3 (Continued)

Simplified Heat Transmission Numbers (HTN)

No.	Concrete	HTN	U
4	4-in. concrete, furred and plastered, no insulation, no flooring above	13.0	0.185
5	Same, with 1-in. rigid insulation	8.8	0.125
(3)	Floors (unheated below)		
	Wood		
1	Joists exposed, double floor, no insulation	5.1	0.07
2	Plasterboard below joists, double floor, no insulation	3.8	0.05
3	Same, but with rock-wool insulation between joists	1.1	0.015
	Concrete		
4	4-in. concrete flooring, no insulation	6.0	0.085
(4)	Floors (on ground)		
	Concrete		
1	4-in. concrete base, no insulation	21.4	0.30
2	Same, with 1-in. rigid insulation	4.4	0.06
3	4-in. concrete, single wood flooring, no insulation	17.5	0.25
4	Same, with 1-in. rigid insulation	3.2	0.045
5	4-in. concrete, double wood flooring, no insulation	5.6	0.08
6	Same, with 1-in. rigid insulation	2.8	0.04
(5)	Floors (above ground)		
	Wood		
1	Joists, double flooring, rock wool, and sheathing	7.0	0.10
	Concrete		
2	4-in. slab, wood flooring, 1-in. rigid insulation	12.0	0.17

TABLE 7-3 (Continued)

Simplified Heat Transmission Numbers (HTN)

No.		HTN	U
(6)	**Roofs**		
	Wood		
1	Shingles, sheathing and plaster, no insulation	21.7	0.31
2	Same, with rock-wool insulation between rafters	8.0	0.11
	Concrete		
3	Concrete slab, roofing and furred ceiling, no insulation	28.0	0.40
4	Same, with 1-in. rigid insulation	12.6	0.18
(7)	**Partitions (unheated)**		
	Frame		
1	Plaster on one side of studs, no insulation	22.0	0.34
2	Same, with rock-wool insulation between studs	8.0	0.11
3	Plaster, both sides, no insulation	12.0	0.17
	Masonry		
4	8-in. cinder block plaster one side, no insulation	15.0	0.21
5	Same, with 1-in. rigid insulation	8.0	0.11
6	4-in. hollow clay tile, plaster one side	16.0	0.228
(8)	**Doors and Windows**		
	Type		
1	Doors, single-glazed windows	79.1	1.13
2	Storm doors and windows	41.0	0.58
3	Double windows (Thermopane)	43.4	0.62

for frame construction with sheathing and wood siding, with the wall not insulated.

Likewise, the HTN factor for single-glazed windows shown in Table 7-3, section 8, may be found by multiplying the heat-loss coefficient U (1.13) by 70°F:

$$\text{HTN} = 1.13 \times 70^\circ = 79.1$$

for single-glazed windows.

Enough typical constructions have been included in the simplified heat-transmission numbers to cover all generally encountered problems.

7-6 WHAT IS INFILTRATION AND WHAT DOES IT HAVE TO DO WITH HEAT LOSS?

Thus far consideration has been given only to heat loss through wall, floors, window glass, and other members by transmission. There is one more consideration. In any room containing windows and doors there is a certain amount of air leakage that seeps inward through the doors and window cracks when the wind blows. This leakage is called infiltration. Every cubic foot of cold outside air that leaks into the room imposes an additional load on the heating system. It is entirely correct to consider the heat required to warm the infiltrated air to room temperature as an additional heat loss, in as much as the heating system has to supply it.

Naturally, infiltration can be lessened by weatherstripping the doors and windows. Storm windows are not very effective in stopping infiltration, but they do aid in cutting down the transmission heat loss through the glass. See Table 7-3, section 8, and compare items 1 and 2.

A fireplace encourages infiltration when it is in use since the air used in creating the draft is sucked in through the cracks around doors and windows.

Another factor to consider when computing infiltration is the number of sides of the room exposed to the weather. A room with one exposed wall will have less infiltration than one with two or three exposed walls. All the above factors have been taken into account in computing the HTN factors for section 9 of Table 7-4.

Note that the HTN factors listed in section 9 should be multiplied by the cubic volume of the room, not by the wall areas.

Cubic content or volume

$$= \text{Room length} \times \text{Room width} \times \text{Ceiling height}$$

When all dimensions are measured in feet, the cubic content is expressed in cubic feet.

TABLE 7-4

Infiltration and Correction Factors

Multiply the cubic volume of the room by factor in the table.

(9) Infiltration

Wind protection	Doors and Windows Weatherstripped		No Weatherstripping	
	One Side Exposed			
	HTN	U	HTN	U
Without fireplace	0.84	0.012	1.4	0.02
With fireplace	2.24	0.032	2.8	0.04
	Two Sides Exposed			
	HTN	U	HTN	U
Without fireplace	1.26	0.018	2.1	0.03
With fireplace	2.66	0.038	3.5	0.05
	Three or Four Sides Exposed			
	HTN	U	HTN	U
Without fireplace	1.68	0.024	2.8	0.04
With fireplace	3.08	0.044	4.2	0.06

(10) Correction Factors

Outside temperature	Inside Temperature (°F)						
	50	55	60	65	70	75	80
40	0.19	0.25	0.33	0.41	0.50	0.60	0.69
30	0.25	0.32	0.40	0.48	0.57	0.67	0.77
20	0.38	0.45	0.54	0.62	0.71	0.81	0.92
10	0.50	0.58	0.67	0.76	0.86	0.96	1.07
0	0.63	0.71	0.80	0.90	1.00	1.11	1.23
−10	0.75	0.84	0.94	1.04	1.14	1.26	1.38
−20	0.88	0.97	1.07	1.17	1.29	1.41	1.54
−30	1.01	1.10	1.21	1.31	1.43	1.55	1.69

7-7 WHAT ARE CORRECTION FACTORS?

Section 10 of Table 7-4 shows the correction factors that, if used as multipliers, will correct any heat-loss computation as figured from sections 1 to 8 of the simplified heat-loss factor table, Table 7-3, from the standard design conditions of 70°F inside and 0°F outside to any other set of design temperatures. Section 10 of Table 7-4 includes a range of inside temperatures from 50 to 80°F, and outside temperatures from minus 30° to plus 30°F. For example, if a structure has a heat loss of 100,000 Btu/hr based on 0°F outside and 70°F inside and is located in Bangor, Maine, then the corrected heating loss is 100,000 × 1.14 or 114,000 Btu/hr.

7-8 HEAT-LOSS DATA RECORDED ON A WORK SHEET

Table 7-5 shows a work sheet that has proved very satisfactory. With its aid, construction changes and their effect on the heat-loss computations are readily evaluated.

Each vertical column on the work sheet has been numbered for quick reference. Column 1 contains the name of the room and its cubic content:

$$\text{Cubic content} = \text{Length} \times \text{Width} \times \text{Height}$$

Columns 2 and 3 describe the heat-loss item being considered and its area. All outside wall, regardless of orientation, is figured at one time for any given room. Next, multiply by the ceiling height to obtain the total area, including windows and doors. To find the net wall area, subtract the sum of the door and window areas from the total wall area. Always figure on the safe side, but do not attempt to be highly accurate. A 10 percent factor of safety is added to each room to allow for slight discrepancies in calculation and uncertainties in construction.

Column 4 shows the heat transmission number (HTN) listed in Table 7-3.

Column 5 is the product of Columns 3 and 4. When figuring infiltration, the HTN is multiplied by the cubic content shown in Column 1.

TABLE 7-5

Sample Work Sheet for Heat-Loss Computations

Name			Design Temperature		Inside____ Outside____	
Location						
1	2	3	4	5	6	7
Room cu ft	Item	Area	HTN	Heat Loss Btu/hr	Correction Factor	Total

Column 6 shows the correction factor for conditions of design other than 0°F outside and 70°F inside. The correction factor may be found in section 10, Table 7-4.

Column 7 is the product of Columns 5 and 6. The heat loss for the entire structure is the sum of the individual room heat losses.

7-9 HEAT-LOSS COMPUTATION FOR A RESIDENCE

EXAMPLE: Calculate the heat loss for the basic house plans shown in Fig. 7-1 and 7-2. Location: Paterson, N.J.
Temperature: Inside, 70°F. Outside, 5°F. All floors above unexcavated spaces to be insulated with rock wool. Ceiling height is 9′-0″.

PROCEDURE: Carefully follow the heat-loss computations shown in Table 7-6. Check the exposed wall dimensions, doors and windows, ceiling dimensions, and cubic content of each room and its infiltration factor, and add a 10 percent factor of safety. Note the correction factor of 0.93. Check, similarly all the figures for the other rooms in Table 7-6 with the dimensions given on the plan for a complete understanding of how a heat-loss computation is made.

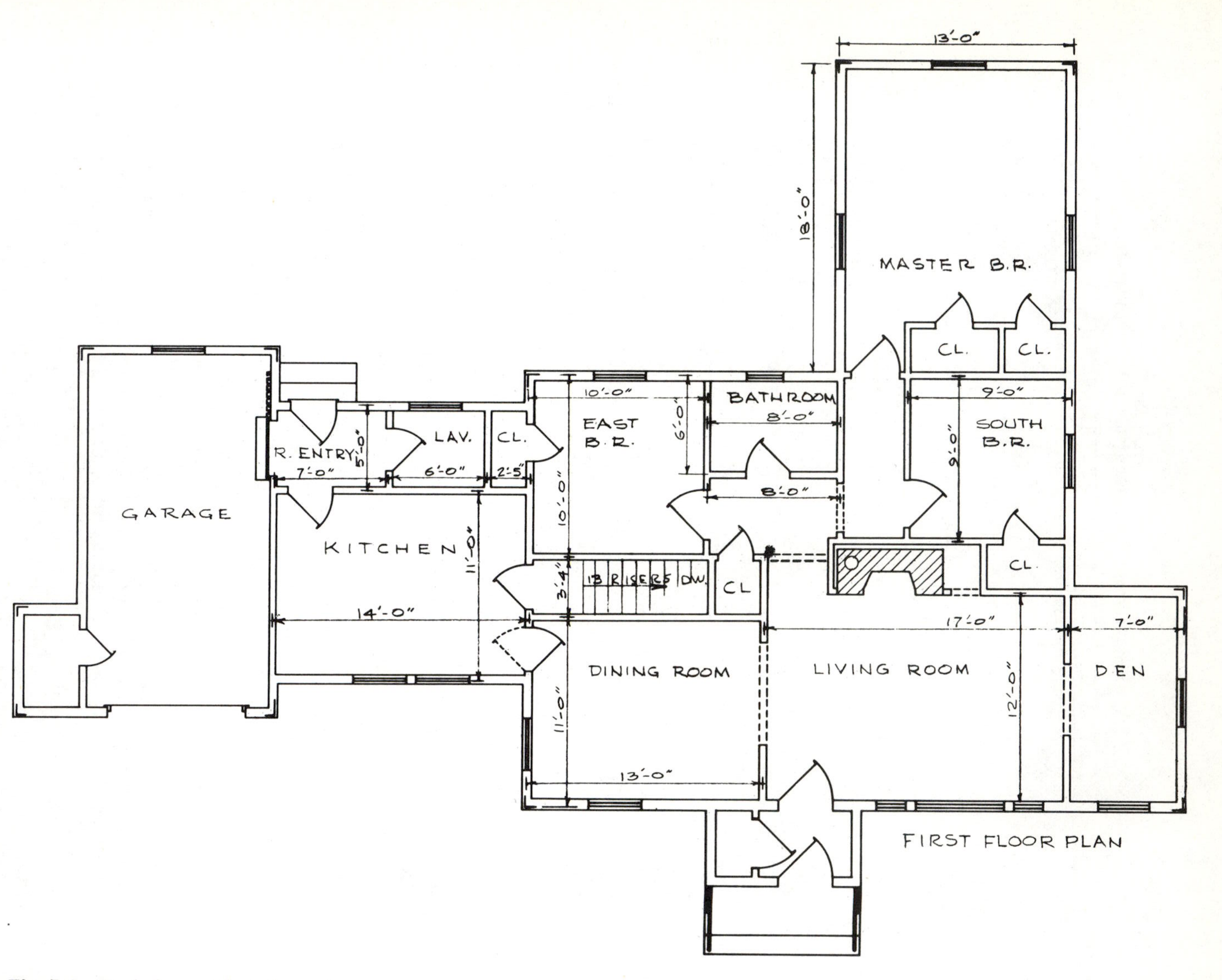

Fig. 7-1 Basic layout first floor plan.

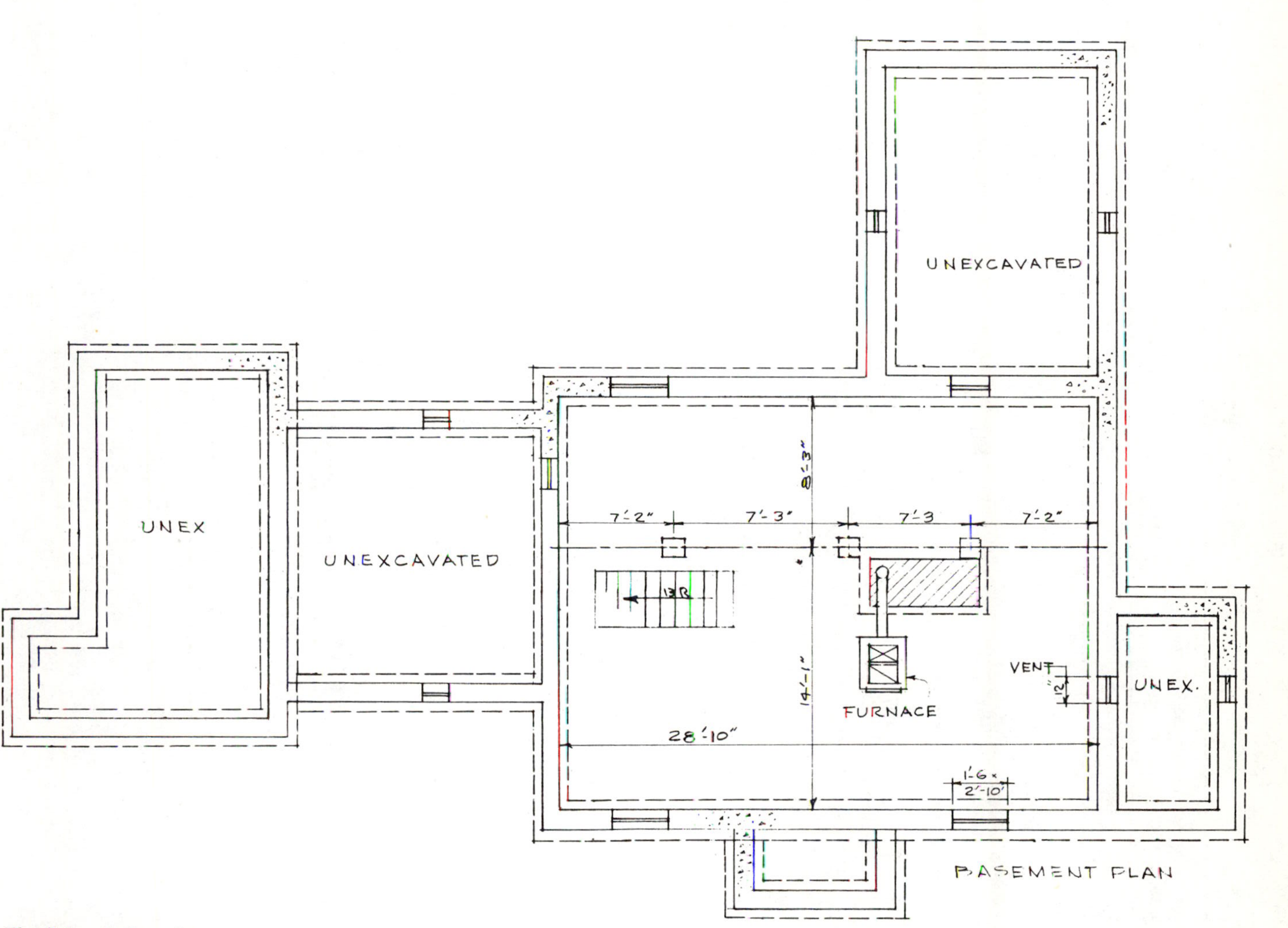

Fig. 7-2 Cellar plan.

TABLE 7-6

Heat-Loss Computation

Name: Frame residence—basic house
Location: Paterson, New Jersey

Design Temperature: Inside 70° F; Outside 5° F

1	2	3	4	5	6	7
Room (cu ft)	Item	Area[b]	HTN	Heat Loss (Btu/hr)	Correction Factor	Total
Living Room 12′ –0″ X 17′ –0″ X 9′ –0″ 1830 cu ft	Exposed wall					
	17′–0″ X 9′–0″	153				
	Door and window					
	20.0 + 34.6	55	79.1	4,350		
	Net Wall	98	18.2	1,780		
	Ceiling 12′ –0″ X 17′ –0″	204	10.5	2,150		
	Infiltration		2.8	5,100		
				13,380		
	Add 10% factor of safety (F.S.)			1,338		
				14,718	0.93	13,700
Dining Room 11′ –0″ X 13′ –0″ X 9′ –0″ 1290 cu ft	Exposed wall					
	20′ –0″ X 9′ –0″	180				
	Windows 12.5 + 12.5	25	79.1	1,980		
	Net Wall	155	18.2	2,820		
	Ceiling 11′ –0″ X 13′ –0″	143	10.5	1,500		
	Infiltration		2.1	2,700		
				9,000		
	Add 10% F.S.			900		
				9,900	0.93	9,200
Kitchen 11′ –0″ X 14′ –0″ X 9′ –0″ 1390 cu ft	Exposed wall					
	14′ –0″ X 9′ –0″	126				
	Windows 9.5 + 9.5	19	79.1	1,500		
	Net Wall	107	18.2	1,950		
	Garage Partition					
	11′ –0″ X 9′ –0″	99	12.0	1,190		
	Floor 11′ –0″ X 14′ –0″	154	7.0	1,080		
	Ceiling 11′ –0″ X 14 ′ –0″	154	10.5	1,620		
	Infiltration		1.4	1,950		
				9,290		
	Add 10% F.S.			929		
				10,219	0.93	9,550
Lavatory 5′ –0″ X 6′ –0″ X 9′ –0″ 270 cu ft	Exposed wall					
	6′ –0″ X 9′ –0″	54				
	Window 2′ –0″ X 3′ –2″	6.3	79.1	500		
	Net Wall	47.7	18.2	870		
	Floor 5′ –0″ X 6′ –0″	30	7.0	210		
	Ceiling 5′ –0″ X 6′ –0″	30	10.5	315		
	Infiltration		1.4	380		
				2,275		
	Add 10% F.S.			227		
				2,502	0.93	2,330

TABLE 7-6 (Continued)

Heat-Loss Computation

1	2	3	4	5	6	7
Room (cu ft)	Item	Area[b]	HTN	Heat Loss (Btu/hr)	Correction Factor	Total
Rear Entry 5′ –0″ X 7′ –0″ X 9′ –0″ 315 cu ft	Exposed wall					
	7′ –0″ X 9′ –0″	63				
	Door – rear					
	2′ –6″ X 6′ –8″	16.6	79.1	1,310		
	Net Wall	46.4	18.2	850		
	Garage Partition	45				
	Fireproof Door	21	39.0	820[a]		
	Net Wall	24	12.0	290		
	Floor 5′ –0″ X 7′ –0″	35	7.0	245		
	Ceiling 5′ –0″ X 7′ –0″	35	10.5	366		
	Infiltration		1.4	443		
				4,324		
	Add 10% F.S.			432		
				4,756	0.93	4,420
Bathroom 6′ –0″ X 8′ –0″ X 9′ –0″ 430 cu ft	Exposed wall					
	8′ –0″ X 9′ –0″	72				
	Window 2′ –0″ X 3′ –2″	6.3	79.1	500		
	Net Wall	65.7	18.2	1,195		
	Ceiling 6′ –0″ X 8′ –0″	48	10.5	505		
	Infiltration		1.4	600		
				2,800		
	Add 10% F.S.			280		
				3,080	0.93	2,860
Den 7′ –0″ X 12′ –0″ X 9′ –0″ 760 cu ft	Exposed wall					
	26′ –0″ N 9′ –0″	234				
	Window 12.5 + 12.5	25	79.1	1,980		
	Net Wall	209	18.2	3,800		
	Floor 7′ –0″ X 12′ –0″	84	7.0	588		
	Ceiling 7′ –0″ X 12′ –0″	84	10.5	885		
	Infiltration		2.8	2,130		
				9,383		
	Add 10% F.S.			938		
				10,321	0.93	9,600
South Bedroom 9′ –0″ X 9′ –0″ X 9′ –0″ 730 cu ft	Exposed wall	8				
	9′ –0″ X 9′ –0″	81				
	Window 3′ –0″ X 4′ –2″	12.5	79.1	990		
	Net Wall	68.5	18.2	1,250		
	Ceiling 9′ –0″ X 9′ –0″	81	10.5	850		
	Infiltration		1.4	1,020		
				4,110		
	Add 10% F.S.			411		
				4,521	0.93	4,200

TABLE 7–6 (Continued)

Heat-Loss Computation

1	2	3	4	5	6	7
Room (cu ft)	Item	Area[b]	HTN	Heat Loss (Btu/hr)	Correction Factor	Total
East Bedroom 10′ –0″ X 10′ –0″ X 9′ –0″ 900 cu ft	Exposed walls					
	12′ –0″ X 9′ –0″	108				
	Window 3′ –0″ X 4′ –2″	12.5	79.1	990		
	Net Wall	85.5	18.2	1,560		
	Ceiling 10′ –0″ X 10′ –0″	100	10.5	1,050		
	Infiltration		1.4	1,260		
				4,860		
	Add 10% F.S.			486		
				5,346	0.93	4,970
Master Bedroom 13′ –0″ X 18′ –0″ X 9′ –0″ 2100 cu ft	Exposed wall					
	49′ –0″ X 9′ –0″	440				
	Windows 12.5 + 12.5 + 12.5	37.5	79.1	2,950		
		402.5	18.2	7,320		
	Floor 13′ –0″ X 18′ –0″	235	7.0	1,640		
	Ceiling 13′ –0″ X 18′ –0″	235	10.5	2,470		
	Infiltration		2.8	5,900		
				20,280		
	Add 10% F.S.			2,028		
				22,308	0.93	20,700

[a]Tabular value cut in half since the temperature difference is 35°F instead of 70°F because room adjacent to fireproof door is unheated.

[b]For computations, use room sizes to the nearest whole foot. Add one foot to ceiling height in computing cubic content and exposed wall area.

7-10 REVIEW QUESTIONS

1. What is the basic rule for computing heat loss?
2. What is the outside design temperature in Columbus, Ohio?
3. Briefly explain the meaning of an HTN factor.
4. How is infiltration calculated?
5. When must a correction factor be used?

8 Heating Systems

8-1 PRACTICAL CONSIDERATIONS

When a house has a basement in which there is an open area or recreation space, the heating unit is usually made a part of that space. Not only are modern heating units designed as basement furniture, but by placing them directly in the open basement area they heat that area effectively. The basement-less house on the other hand has its heating unit in the attached garage or in a closet-like space in the living area.

It is considered good practice to place the heating unit as near the chimney as possible because long smoke pipes reduce the efficiency of the unit.

The heating system of any home assumes more and more importance when it is realized just how much of a part it plays in the success of a well planned project. The design of the heating system may directly affect the following:

1. The shape, size, and location of windows, if radiators are to be located beneath them.
2. Location of the recreation room in basement, to include heating unit; heating unit in ventilated closet or garage where no basement exists.

3. Location of the chimney.
4. Thickness, stud spacing, and location of partition walls where warm-air ducts are used.
5. Construction of floors and ceilings where radiant heat is contemplated.
6. Location of doors to provide convenient wall spaces for warm-air return registers.
7. Construction of exterior walls to achieve good heat insulation.
8. Physical shape of house to ensure compactness and consequent economy of heating where necessary.

Figures 8-1 through 8-17 show how the various types of heating systems function. Advantages and disadvantages are given for each system and the type of building each may be used best in are discussed.

8-2 FORCED WARM-AIR DUCT SYSTEMS

The forced warm-air duct system, Fig. 8-1, is perhaps the most common heating system in use today. It finds its greatest use in residential and small public structures but is commonly found in offices and other types of buildings.

It meets all the requirements of winter heating, air cleaning, and circulation. It can easily be converted into a year-round air conditioning system by the addition of summer cooling equipment. The same duct work is used.

8-3 PLACEMENT OF SUPPLY AND RETURN AIR DIFFUSERS

Air diffusers (Fig. 8-2) are round, square, or rectangular metal outlets that deliver warm or cool air. They may be fixed to ceiling, wall, or floors and are attached to a duct system. Air diffusers may be either grilles or registers. Registers are placed at the ends of ducts leading into a room and are provided with adjustable louvers so that the air can be blown into a room at different angles. They can also be partially closed or entirely.

Grilles are like registers except they do not have adjustable louvers. They have crossed parallel bars and are usually fixed in position.

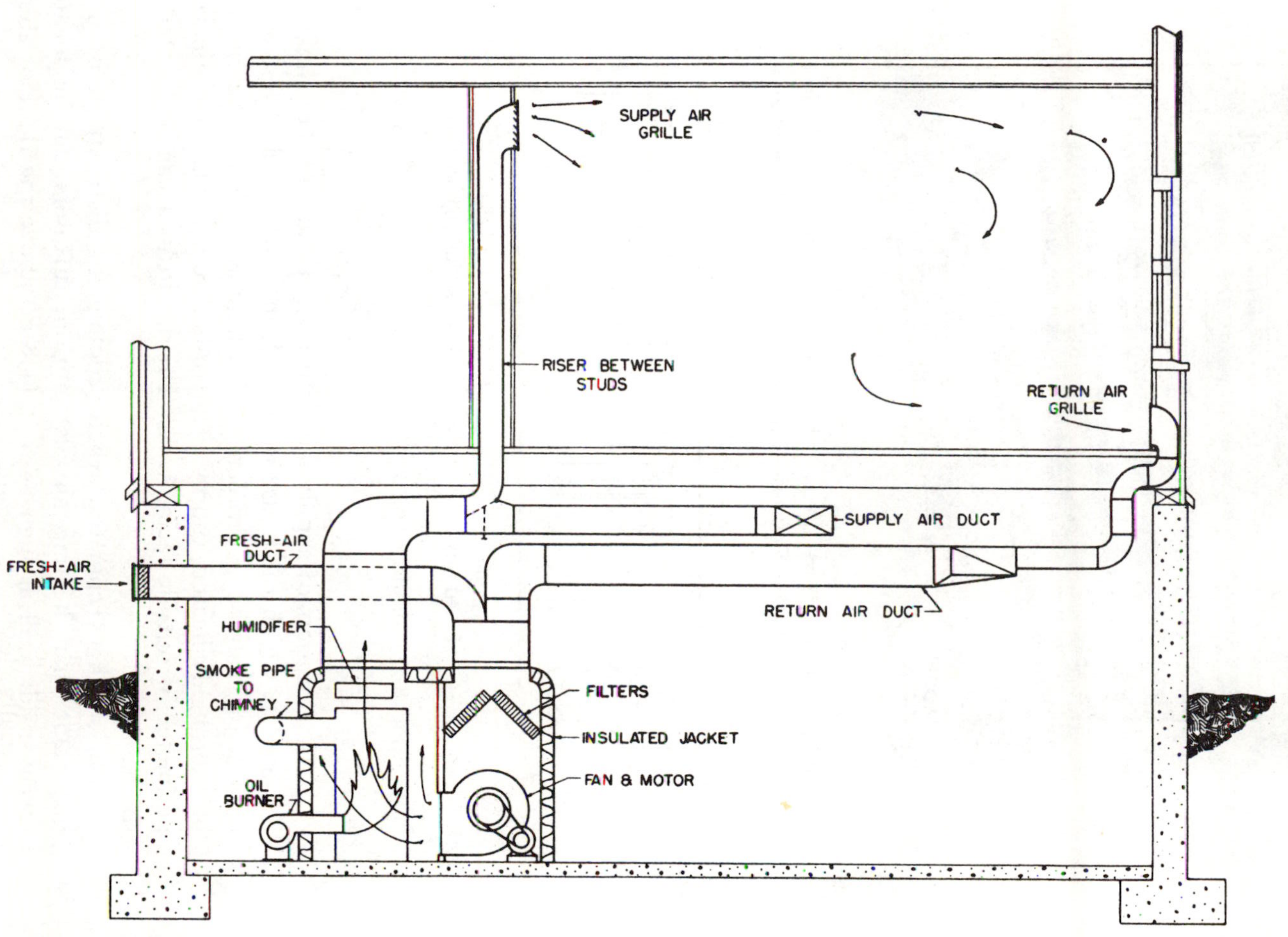

Fig. 8-1 Forced warm-air duct system. Duct arrangement is ideal for heating and cooling.

Adjustable air diffuser used in a heating and/or cooling supply air distribution system. Courtesy Environmental Elements Corporation.

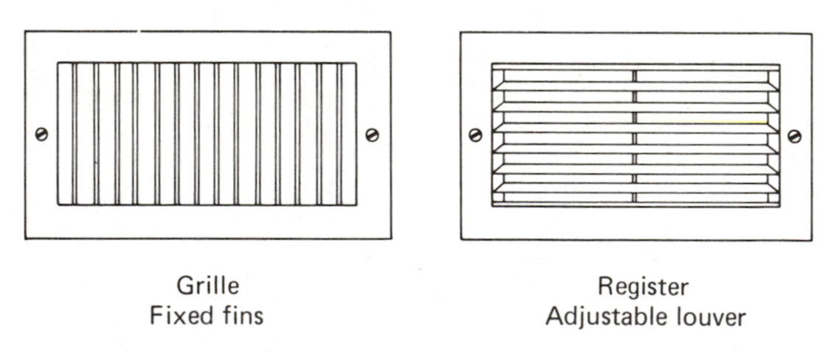

Fig. 8-2 Air diffusers.

The location of supply air diffusers depends on the climate of the area in which the house is located. In northern climates, where cold floors can be a problem and where outside exposures must be thoroughly heated, supply air diffusers are best placed around the perimeter of outside facing walls. The air diffuser can be located in the baseboard.

In southern climates, where cooling air distribution is more important than heating air distribution, diffusers can be located on an inside wall as high as 6 ft (1.8 m) above the floor or about 6 in. (152 mm) from the ceiling.

8-4 PLACEMENT OF RETURN AIR GRILLES

In northern climates, where supply air diffusers are located along the perimeter, return air grilles are usually placed near the baseboard on inside walls. They may also be placed high on the inside walls to prevent drafts.

In southern climates, where high inside wall outlets or ceiling diffusers are used, returns may be placed in any convenient location on inside walls. Where one or more supply outlets are placed in every room to be heated, connecting rooms with open doorways can share a common return that may be located in a hallway or foyer. Garages, kitchens, and bathrooms do not require returns.

Where cooling as well as heating is supplied from the same outlet, it is necessary to direct the supply air upward for cooling. Cold air is heavier than warm air and tends to collect near the floor. To raise cold air, baseboard diffusers should have adjustable vanes or baffles.

As a rule no special construction is required where ducts run through partitions as the ducts are made to fit the standard stud space. Thus, a 3¼ by 14 in. (82 by 356 mm) rectangular duct is about the largest size duct that can be run between two studs spaced 16 in. (406 mm) on center. For rooms larger than 10 ft by 10 ft (3 m by 3 m) two or more ducts and outlet diffusers are required. A warm air system is only as good as its distribution system, and therefore, it is important to provide for sufficient supply and return air openings.

The furnace unit does not have to be located in the center of the house, and may be placed wherever convenient. Ducts may be run in the basement or in an attic space, or along the ceiling of the space being heated. If the ducts are run along the ceiling, air outlets of special design may be mounted on the ceiling. A typical gas-fired, warm-air furnace, including an air conditioner, air cleaner, flue vent damper, and humidifier is shown in Fig. 8-3.

The advantages of the forced warm-air duct system can be stated as follows:

1. Provides all the requirements of winter air conditioning, air cleaning, and air circulation.
2. May be converted into year-round air conditioning by adding summer cooling equipment.

1. Condensing unit
1A. Heat pump
2. Electronic air cleaner
3. Humidifier

Fig. 8-3 Forced warm-air furnace, gas-fired; including air conditioner, air cleaner, flue vent damper and humidifier. (Reproduced with permission of Carrier Corporation, © copyright 1979, Carrier Corporation.)

3. System cannot freeze in the winter.
4. Can circulate air without cooling nor heating.

8-5 RADIAL DUCT SYSTEMS

The radial duct system (Fig. 8-4) consists of a series of pipes running from the furnace, usually centrally located, to supply air outlets located around outside walls of the various rooms. The centrally located furnace causes the duct runouts to be of about the same length, a necessary requirement for uniform distribution. This system may be constructed in the concrete floor slab or in a crawl space. Supply air outlets may be placed in the floor and under windows.

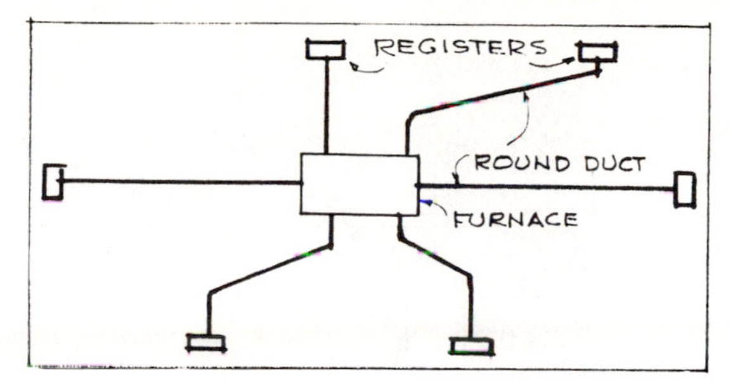

Fig. 8-4 Radial perimeter duct system.

8-6 PERIMETER LOOP DUCT SYSTEMS

The perimeter loop system (Fig. 8-5) is somewhat similar to the radial system, except that a single duct running around the perimeter or outer edge of the house supplies air to each diffuser in the floor of the various rooms above. The runout ducts from the furnace to the perimeter loop are generally larger than those on the radial system and fewer such ducts are therefore needed. This system supplies extra heat along the perimeter of the house at the floor and the outside walls where it is most needed. The greatest heat loss occurs along the perimeter.

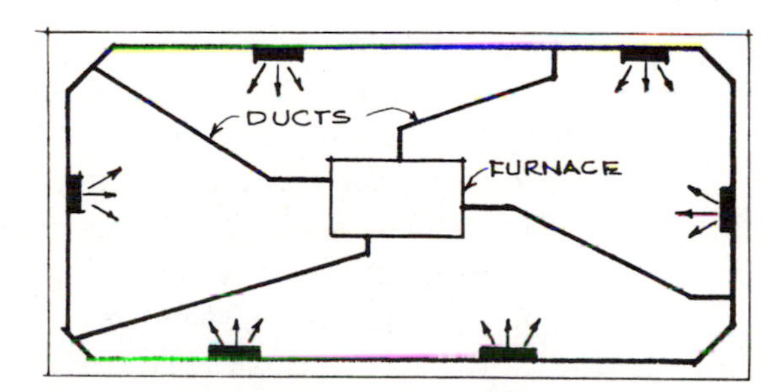

Fig. 8-5 Perimeter loop duct system in concrete floor slab.

8-7 ONE-PIPE FORCED HOT WATER HEATING

The most popular type of hot water radiator system for use in residential work is the one-pipe forced hot water system, shown in Fig. 8-6. In a one-pipe layout, a single main, usually in the base-

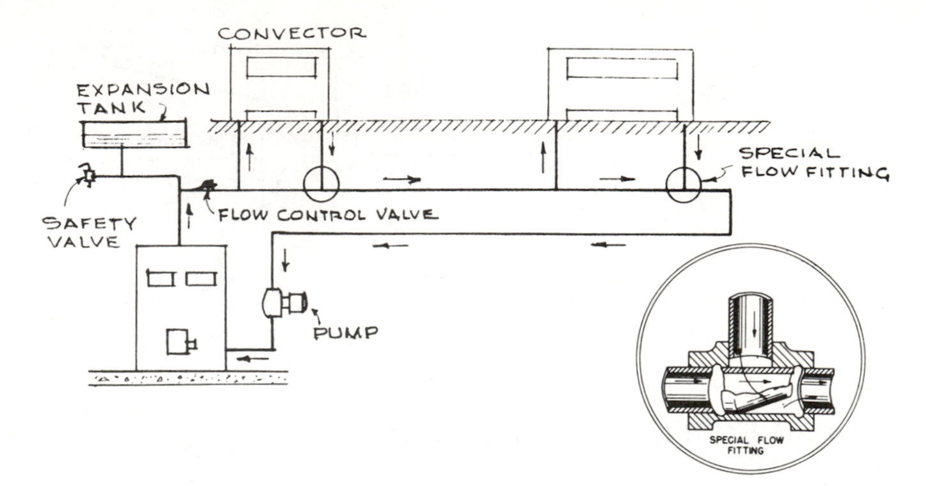

Fig. 8-6 One-pipe forced hot water system.

ment, follows the perimeter of the building, and branches and risers connect with individual radiators.

The piping for the one-pipe hot water system may be wrought iron, black steel, or copper pipe. Joints may be screwed, welded, or sweat-soldered in the case of copper pipe. The size of the main as it leaves the boiler is usually 1¼ or 1½ in. (32 or 38 mm) in average residential work, and the size of the radiator branches to the first floor are ½ in. (12.7 mm). Risers to the second floor seldom exceed ¾ in. (19 mm).

The average pump on the one-pipe system pumps about 8 to 10 gal (30 to 38 *l*) of hot water per minute through the system and requires a ⅙-hp (0.124-kW), 110-V motor.

A flow control valve is located at the first elbow on the main leaving the boiler. The purpose of this valve is to open under the influence of the pressure of the flowing water when the pump is operating, and to close and prevent gravity flow when the pump is off. If the hot water was allowed to circulate during pump off periods, the radiators would overheat when the room was already too warm.

A special flow fitting is also installed at the point where the water leaves the convector and joins the supply water to the next convector. The cooler water leaving the convector flows in the same pipe with hotter water feeding the next convector. The special flow fitting is designed to lead the cooler and hotter water together in the same pipe without undue mixing.

Radiators may be of the familiar cast-iron thin-tube design, or may be the more modern convector, or baseboard units with fins. The advantage of the convector lies in its streamlined appearance, and in the fact that it may be recessed in the wall. Each convector or radiator is usually equipped with a control valve and air vent.

Whenever possible, place radiators beneath windows or against cold walls. The best location is underneath a window because cold air leaking through the window is warmed before it reaches the interior of the room or floor.

If large picture windows extending nearly to the floor are used, it may not be possible to place a radiator beneath the sill. Under these circumstances a forced warm-air system or some form of radiant heat (such as a radiant baseboard) might prove advisable. Sometimes double-glazed windows, Thermopane for example, are used to make the heating less of a problem.

In operation, the one-pipe hot water system is filled slowly with water, with air vents open, until all air is vented from the system. Then the vents are closed and the control valves adjusted, and the system is ready for continuous operation.

An aquastat, located in the boiler, continually controls the burner, maintaining boiler water at the desired temperature, usually 180 to 200°F (82.2 to 93.3° C). A room thermostat, sensing a falling temperature in the room, starts the pump that circulates the hot water throughout the system. When the room has warmed to the desired temperature, the thermostat stops the pump and the flow valve swings shut, and the system furnishes no further heat until called upon again by the room thermostat.

The one-pipe hot water system shares top place in popularity today with the forced warm-air system.

Where used:

1. Residential structures
2. Apartment buildings
3. Stores
4. Factories
5. Small public buildings

Advantages:

1. Extremely efficient because of instant circulation of water.
2. Low-tempeature heat for mild weather.
3. Radiators hold heat for a long time.
4. One or more radiators or convectors can be shut off without interfering with the flow of water to other radiators or convectors.

Disadvantages:

1. System always full of water. In cold climates the system must be drained when not in use.
2. May overheat when weather turns mild suddenly.

8-8 EXPANSION TANKS

Expansion tanks are used in hot water systems to do two things:

1. Provide space for expansion of the hot water in the system.
2. Trap air above their water level and compress it to an extent where all the water in the system will be under sufficient pressure at all times to permit relatively high water temperature to be carried in the boiler without the danger of steam being formed. It is on account of this latter function that these tanks are often called compression tanks. See Fig. 8-7.

Although water boils at 212°F (100°C) under atmospheric pressure, its boiling point may be raised to 270°F (132.5°C) when the air pressure in the expansion tank is 30 psi (206.8 kPa). To obtain 215°F (101.7°C) water in the boiler without danger of boiling, which is necessary in order to have 200°F (93.3°C) average water temperature in the radiators, a pressure of at least 12 psi (82.7 kPa) should be carried in the expansion tank. By elevating the tank above the boiler, the boiling point of the water can be increased simply by the increased head. Expansion tanks for average installation require 18- to 24-gal (68- to 91-*l*) tank capacities.

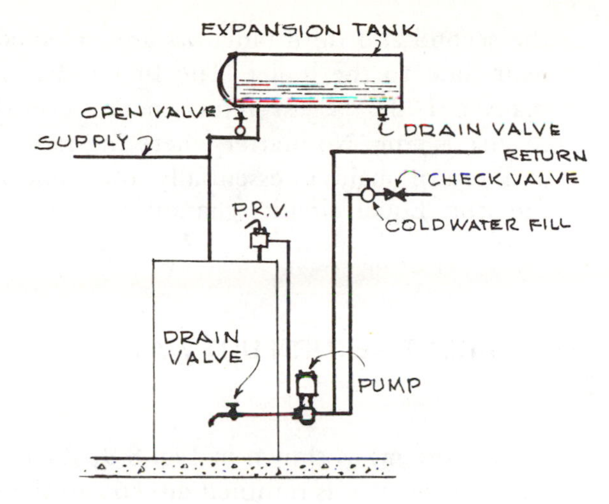

Fig. 8-7 Expansion tank.

8-9 TWO-PIPE REVERSED RETURN HOT WATER SYSTEM

The two-pipe reversed return hot water system, Fig. 8-8, creates circuits from the boiler to radiators of about equal lengths. The hot water coming from the boiler into the first radiator also feeds the second and third and so on. The return line coming from the first radiator, containing somewhat cooler water, and the return line from

Fig. 8-8 Two-pipe reversed return hot water system.

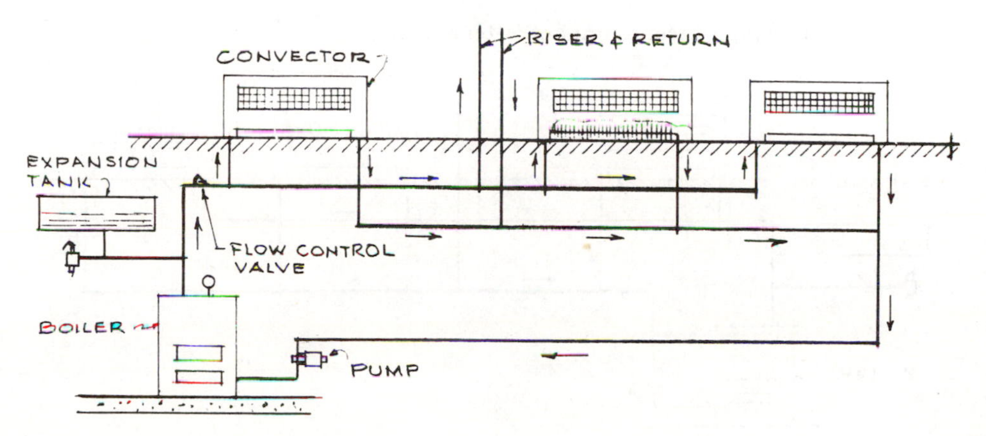

the second and third radiators are collected and brought back in one main line to the boiler. The first radiator has the shortest supply main but the largest return main. For the farthest radiator, the reverse is true. No matter where the radiator is positioned its supply and return main is essentially the same length as for every other radiator. Equal circuits guarantee a delivery of an equal amount of heat.

8-10 TWO-PIPE DIRECT RETURN HOT WATER SYSTEM

In this sytem, as shown in Fig. 8-9, the hot water from the boiler to the first radiator is returned directly to the boiler, making a relatively short circuit of supply and return. The second, third, or the farthest radiator from the boiler, each have a separate return to the boiler making each circuit of supply and return different in length. It is obvious that since the first radiator has a shorter circuit it will get more hot water and will be hotter than the last radiator with the longest circuit.

In order to correct this condition, balancing cocks can be installed, or the proper pipe sizes can be used, or the radiators that receive a greater amount of heat can be proportionately reduced.

EXAMPLE: Count all the radiators except the first one on the circuit. Divide this number into 10 percent (0.10). The answer is the percentage to add to each radiator capacity so that the last radiator on the circuit will have its capacity

Fig. 8-9 Two-pipe direct return hot water system.

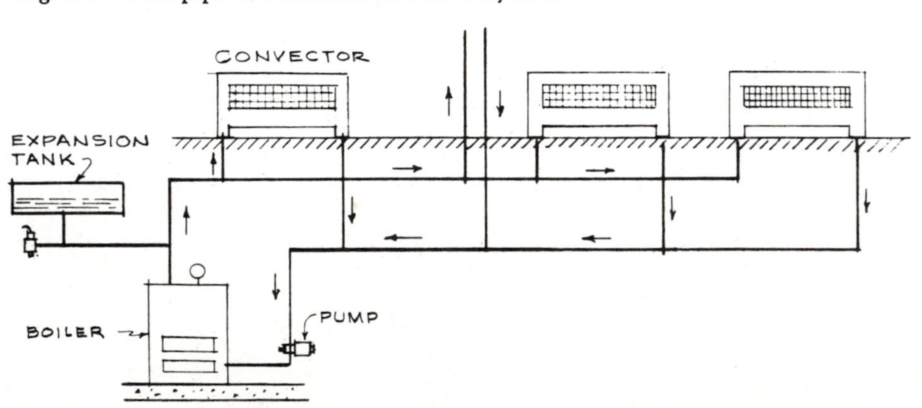

increased by about 10 percent. Note that when 2 percent is added to a number, the number is multiplied by 1.02.

8-11 AIR CONDITIONING SYSTEMS

The air conditioning system (Fig. 8-10) for a small office building is provided with heating and cooling coils for winter and summer air conditioning. Air passes through a spray humidifier, heating or cooling coil, and filters, screening out dust and other particles. Volume dampers control the amount of outside air intake.

8-12 VENTILATING SYSTEMS

The ventilating system (Fig. 8-11) has heating coils for winter heating. This system is used also for ventilating when the heating coils are not in use. Ventilating is the supply and removal of air to or from any space. This may be done by natural or mechanical means.

Natural ventilating is ventilation through open windows and louvered doors. Building codes call for openings of 5 percent of the floor area for each room or habitable space. A room 12 ft by 14 ft (3.7 × 4.3 m) has an area of 168 sq ft (15.6 m^3). Five percent of 168 is 8.4 sq ft (0.78 m^3). This amount is the clear minimum opening needed for natural ventilation.

Window locations play an important role in natural ventilation. Good cross ventilation calls for windows on two opposite ends of two walls. Natural ventilation improves when room doors are left open.

Ventilating the attic space is a matter of providing louvered openings at both ends of a gable roof. The size of the louvered opening for an attic space is 1/300th of the attic floor area. An attic floor measuring 24 ft by 40 ft (7.3 × 12.2 m) has 960 sq ft (89.2 m^2). The louver should be 1/300th of 960 or 3.2 sq ft (0.3 m^3). To change the 3.2 sq ft into square inches multiply by 144 (0.0929 m^2). Then 3.2 times 144 is approximately 460 sq in. Each end will take half of 460 or 230 sq in. (1484 cm^2). A 14 in. by 24 in. (3.56 × 6.10 m) louver will be about the right size.

Mechanical ventilation can be achieved through the use of fans and blowers. It is first necessary to find the total cubic volume of the air space to be ventilated. Second, using the suggested air change

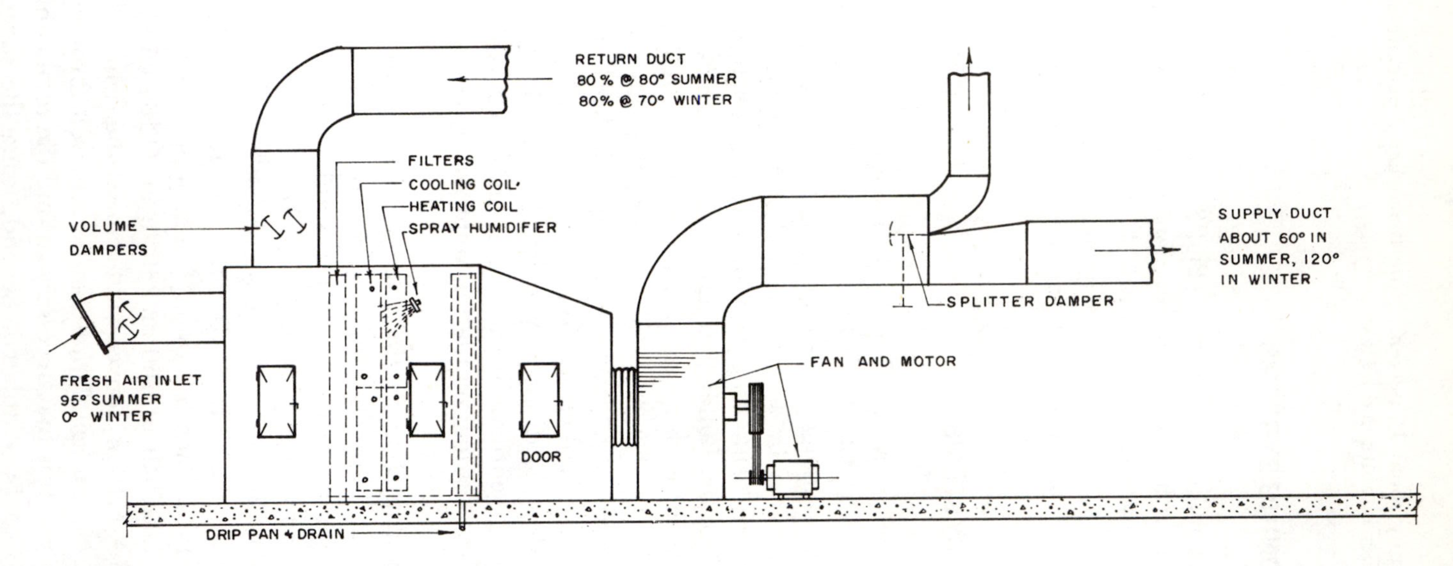

Fig. 8-10 Air conditioning system.

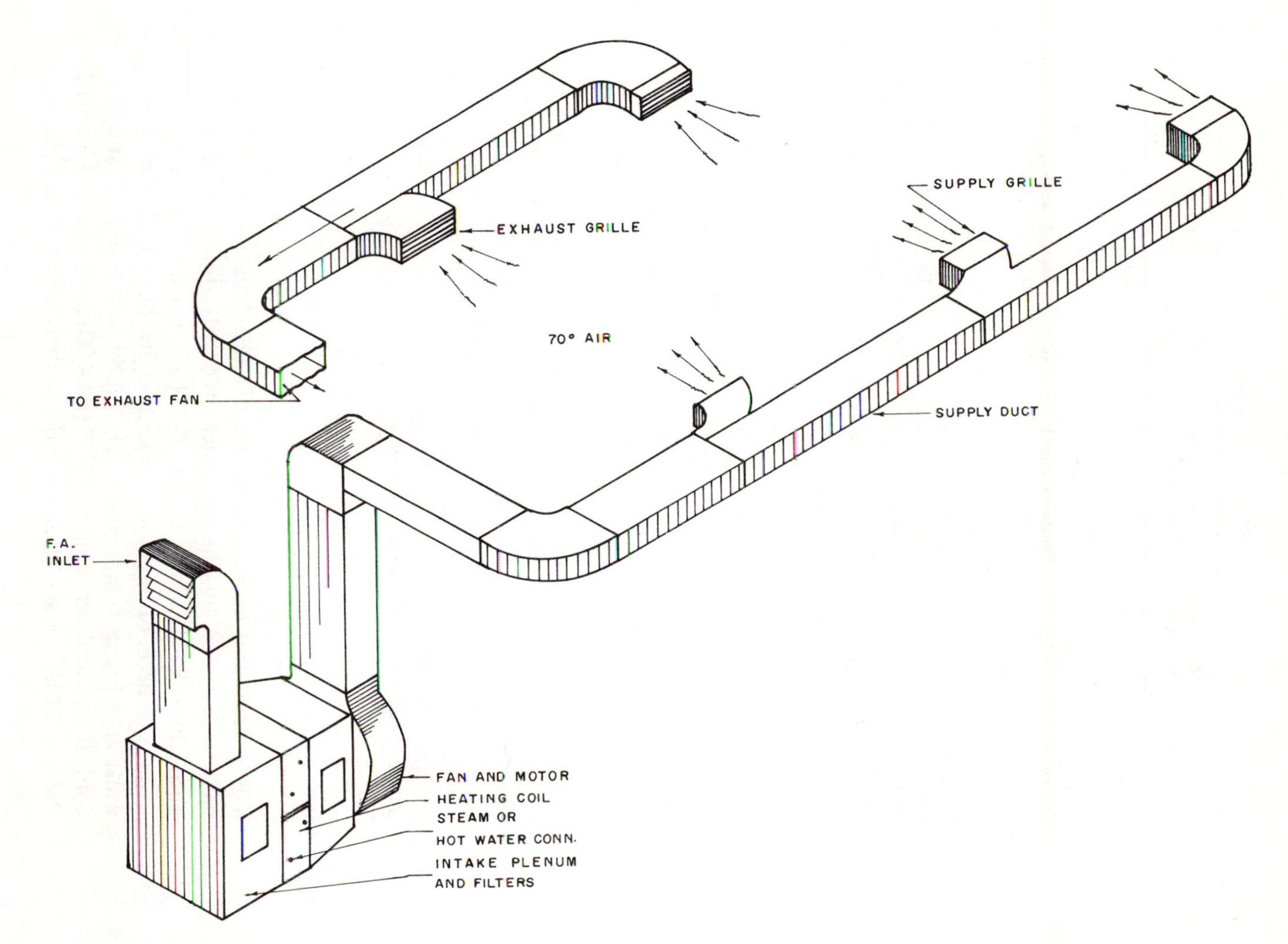

Fig. 8-11 Ventilating system.

TABLE 8-1

Suggested Air Change Rates

Type of Building	Minutes per Air Change	Type of Building	Minutes per Air Change	Type of Building	Minutes per Air Change
Assembly hall	3-10	Dining room	10-15	Mill, paper	2-8
Attic	2-4	Engine room	1-2	Mill, textile	5-15
Auditorium	4-15	Factory (light)	5-10	Offices	2-8
Bakery	1-5	Factory (heavy)	1-6	Packing house	2-5
Bank	3-1	Forge shop	1-2	Projection room	1-2
Bar	2-5	Foundry	1-3	Recreation room	2-8
Beauty parlor	4-10	Garage	2-10	Residence	2-6
Boiler room	1-4	Generating room	2-5	Restaurant	5-10
Bowling alley	2-8	Glass plant	1-2	Retail store	3-10
Church	4-10	Gymnasium	2-10	Theater	3-8
Class room	4-6	Kitchen	1-3	Toilet	2-5

Courtesy of Western Engineering & Mfg. Co.

rate chart shown in Table 8-1, find the best selection which meets the conditions of the air space to be ventilated.

$$\text{CFM} = \frac{\text{Air space volume in cubic feet}}{\text{Air change rate}}$$

EXAMPLE: A building which is 50-ft wide, 50-ft long and 20-ft high, is used as an engine room. Find the required CFM to properly ventilate this building. First, find the volume of the air space by multiplying 50 ft $\times$ 50 ft $\times$ 20 ft, which equals 50,000 cu ft. Referring to the suggested air change rate (Table 8-1), it can be seen that the air change rate for an engine room is a 1 to 2 min air change. Using the above formula, the required CFM would be 25,000 CFM. A fan size with this capability can be selected.

8-13 RADIANT HEATING

The principle of heating by means of radiant heat is usually not very well understood by the beginner. We will therefore, discuss it in nontechnical language. Figure 8-12(A), shows that heat radiates from a source of high temperature (the body) to a receiver (the wall) at a lower temperature. The body will lose heat and become cold. If the wall is heated to 85°F (29.4°C) as shown in Fig. 8-12(B), the rate of heat transferred from the body is lessened, and just enough

Fig. 8-12 Radiant heat.

heat is radiated from the body to offset the metabolic rate of the individual.

Thus, the body feels comfortable. If the wall is heated further as in Fig. 8-12(C), the body actually receives heat and becomes uncomfortable, since it cannot now dissipate its own heat. In convection and air-current heating the body is surrounded with warm air, but in radiant heating, the body is surrounded with warm surfaces. The air temperature between the body and the wall is a secondary consideration, and a seated person may feel entirely comfortable in a temperature of 58°F (14.4°C), as long as the walls, ceiling, floors, and furniture are at the correct temperature to prevent excessive body heat loss.

In radiant heating, no air is introduced, no radiators are used, and consequently few convection currents are set in motion. The heat rays travel in straight lines between the heat emitter (wall, floor, or ceiling panel) and any other surface that is cooler.

Since there are many surfaces that act as reflecting surfaces within the room itself, there exists an endless pattern of radiant rays traveling in all directions. When the surfaces of the room and the furnishings have warmed up to the proper temperature, comfort exists.

Temperature variation within a floor-panel radiant heated room is shown in Fig. 8-13. Warm water is circulated through pipes embedded in the concrete floor. The entire floor becomes warm, emitting heat from a 70°F (21.2°C) floor surface.

The radiator in a convection system also emits a certain amount of radiant rays. Its small surface, at a high temperature of 200°F (93.3°C) however, is more efficient as a producer of convection heat than as a source of radiant heat. The small radiator contains as much heat as the large warm floor or ceiling, but the floor or ceiling presents a flat surface ideal for emitting rays of heat at low temperatures. No air can flow through the floor as it flows through the hot sections of the radiator, and consquently convection currents are reduced. When radiant heat coils are located in the ceiling, there is maximum reduction of convection currents, as the layer of warm air directly under the ceiling cannot rise any higher.

You may wonder which principle of heating is best, and which type of system utilizing that principle is best. There is no simple answer to this question, because too many variables are involved. A large factor is the human one. Not all people agree on the extent of their comfort or discomfort. Some dislike steam systems and

Fig. 8-13 Temperature variations.

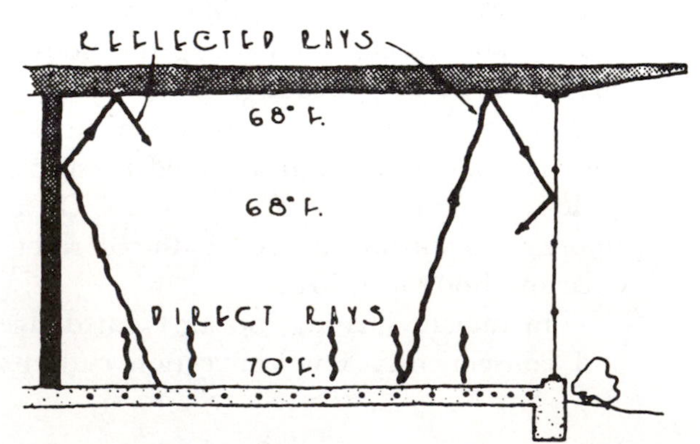

others swear by them. There are those who think radiant heat is too new to be proved as yet. Others think they must have a concrete floor if they want radiant heat.

8-14 RADIANT HEATING SYSTEMS

Radiant heating systems have emerged from a long period of experimentation, improvement, and public education. Thousands are used throughout the world, and simple methods of calculation have been devised.

All the previously described systems utilize any type of fuel desired, such as oil, electricity, or gas. The choice of fuel depends on location, cost, availability, and of course, the user's preference. Automatic control devices are an absolute necessity if even temperatures and high efficiency are to be maintained.

8-15 RADIANT HEATING COILS IN THE CEILING

Figure 8-14 shows a typical room with a radiant heating panel located in ceiling. Hot water is circulated through the ceiling coil, which is embedded in plaster.

Figure 8-15 shows a radiant heating system that consists of serpentine coil pipe that can be embedded in a concrete floor slab.

8-16 DETAILS FOR RADIANT HEATING

The main parts of the radiant heating system are the boiler, automatic firing device (oil, gas, or electric burner), expansion tank, pump or circulator, flow control valve, embedded heating coils, and automatic controls.

In operation, the system is completely filled with city water through an automatic feed valve. Air is vented through vent connections located at the high point of the system and at the expansion tank. An outdoor-indoor control device regulates the boiler water temperature in response to the difference between indoor and outdoor temperatures. As the weather grows colder, the boiler water temperature is raised, and vice versa. The pump, usually driven

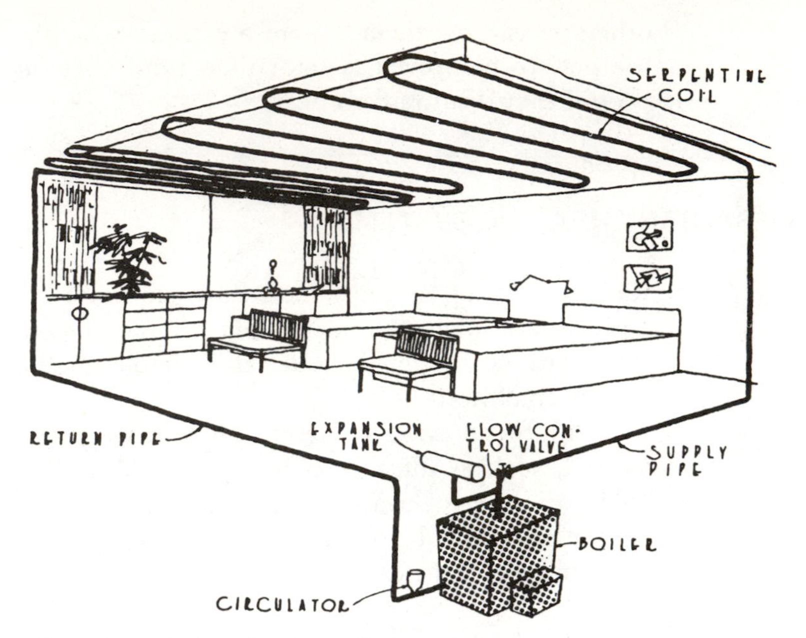

Fig. 8-14 Radiant heating coils in ceiling.

Fig. 8-15 Radiant heating coils in floor.

by a ⅙-hp (0.124-kW), single-phase, 110-V motor, starts and stops in response to a room thermostat.

The function of the heating coils, whether located in the floor or ceiling, is to raise the average temperature of all surfaces enclosing the room, as well as those within the room (such as furniture, pictures, built-in cabinets) to a temperature sufficient to allow normal radiation of body heat. If the average surface temperature is above 70°F (21.1°C) the room air temperature may be below 70°F, to produce conditions of comfort.

Rooms with large glass areas usually require that the heating coils be placed in the ceiling, because panels can safely be heated to about 120°F (48.9°C) and thus more heat may be radiated to the other surfaces.

Floor panels, on the other hand, are usually limited to a temperature of 85°F (29.4°C) in areas in constant use, and to 90°F (32.2°C) along the perimeter of rooms with exposed walls. Sometimes a combination of floor and ceiling panels are used. Less popular in small house construction is the wall panel. Wall temperatures are usually limited to 100°F (37.8°C).

Piping materials are either copper, wrought iron, or black steel. Coils may be welded if steel, or sweated if copper. Screwed fittings are not usually used. The pipe coils may be grid-type, serpentine, or continuous square coil as shown in Fig. 8-16. Pipe sizes vary from ⅜ in. (9.5 mm) for copper tubing used in ceiling panels, to 1¼-in. (31.75-mm) pipe used in floor slabs. Copper tubing used in floor slabs is usually ½ in. (12.7 mm), ¾ in. (19 mm) or 1 in. (25.4 mm), and may be formed into bends with 3- to 6-in. (76- to 152-mm) radii, depending on the pipe diameter.

8-17 RADIANT HEATING CEILING PANELS

The average water temperature in the coils is usually between 100 and 150°F (37.8 and 65.6°C) with the higher values used in ceiling coils. With such low water temperatures, there is no danger of cracking either concrete slab or a plaster wall or ceiling, since the rates of expansion for copper, steel, and wrought iron are practically identical with those of concrete and plaster.

Concrete slab floors need not be bare to produce good results. They may be floored with cemented wood blocks, linoleum, asphalt or vinyl tile, regular wood flooring, or simply carpeted. The effect of floor covering that tends to insulate the floor is to require a slightly higher operating water temperature.

The first cost of a radiant heating system compares favorably with other types of heating systems, and the operating costs have

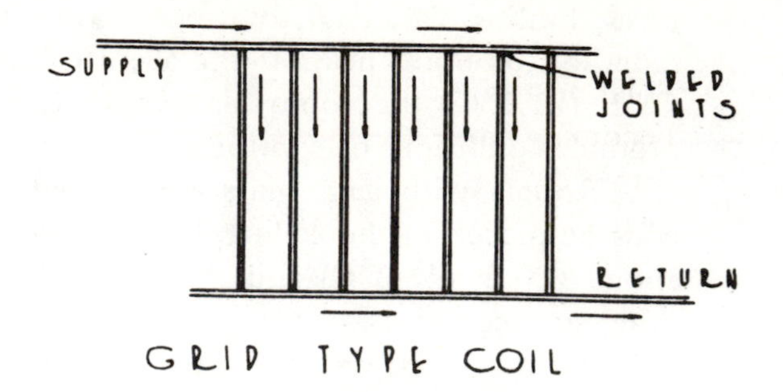

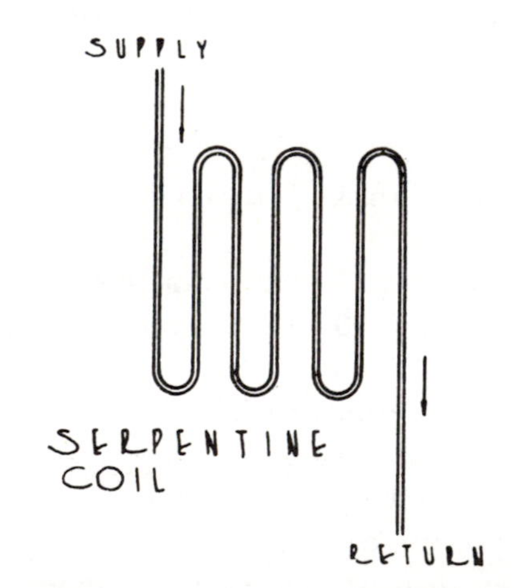

Fig. 8-16 Types of radiant heating pipe coils.

been proven to be as much as 30 percent less than that for more familiar radiator and warm air systems. Figure 8-17 (pp. 162-3) shows radiant heating piping installed in ceilings and floors.

8-18 SOLAR HEATING

Solar heating makes use of the sun's rays to heat the house. This is becoming more and more important because of our decreasing fuel supply. We are again searching for a new source of energy. The sun's heat is one of the obvious answers. There are many questions and problems to be solved before we can harness the sun. Some progress, however, has been made. Recently, a house of 1600 sq ft (149 m^2) was designed that used solar energy to supply about 80 percent of its heating needs and all of its cooling.

Winter heating was supplied by a bank of water filled drums along the house's south face. Warmed by the sun to temperatures in excess of 100°F, the drums act as radiators to heat air drawn around them and inside the house.

The heat is then distributed through the house by a series of ducts and floor registers. When the drums do not meet the full requirements of the house, a conventional furnace supplies the difference.

Summer cooling is provided by a large bed of rocks underneath the house. Air is drawn over the rocks into the house and eventually out to ventilators in the ceiling.

Other designed homes draw sunlight through specially aligned windows to heat tile floors and a series of decorative columns. The columns, filled with water, will accumulate heat during the day and release it at night, supplying about half the winter heat requirements for the house.

Among possibilities, designers are experimenting with the use of cool air that exists in basements or above ground foundations, with placing houses partially underground, and with greenhouses.

In one design a greenhouse was built along the length of the south side of a house. In the winter, the sun heats the air inside the greenhouse and funnels it through vents into second-story rooms. More important, the sun shines directly on a series of water-filled barrels that store the heat for evening hours providing about 40 percent of the house's heating needs.

In addition to raising customary crops, a greenhouse can produce a crop of warm air as shown in the house as shown in Fig. 8-18. The schematic drawing shows the techniques that are used for both winter heating and summer cooling. This installation is a good example of the way a house can be remodeled to take advantage of the benefits of solar heating and cooling.

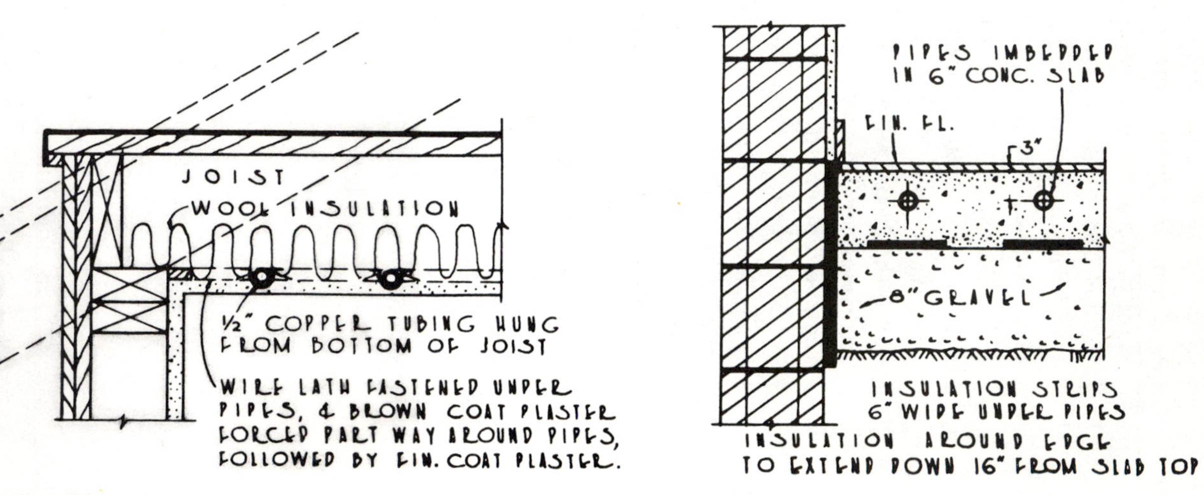

FLAT OR PITCHED ROOF CONSTRUCTION
FLAT OR PITCHED ROOF CONSTRUCTION

A GOOD CONSTRUCTION FOR CONC. FL. SLAB ON GROUND

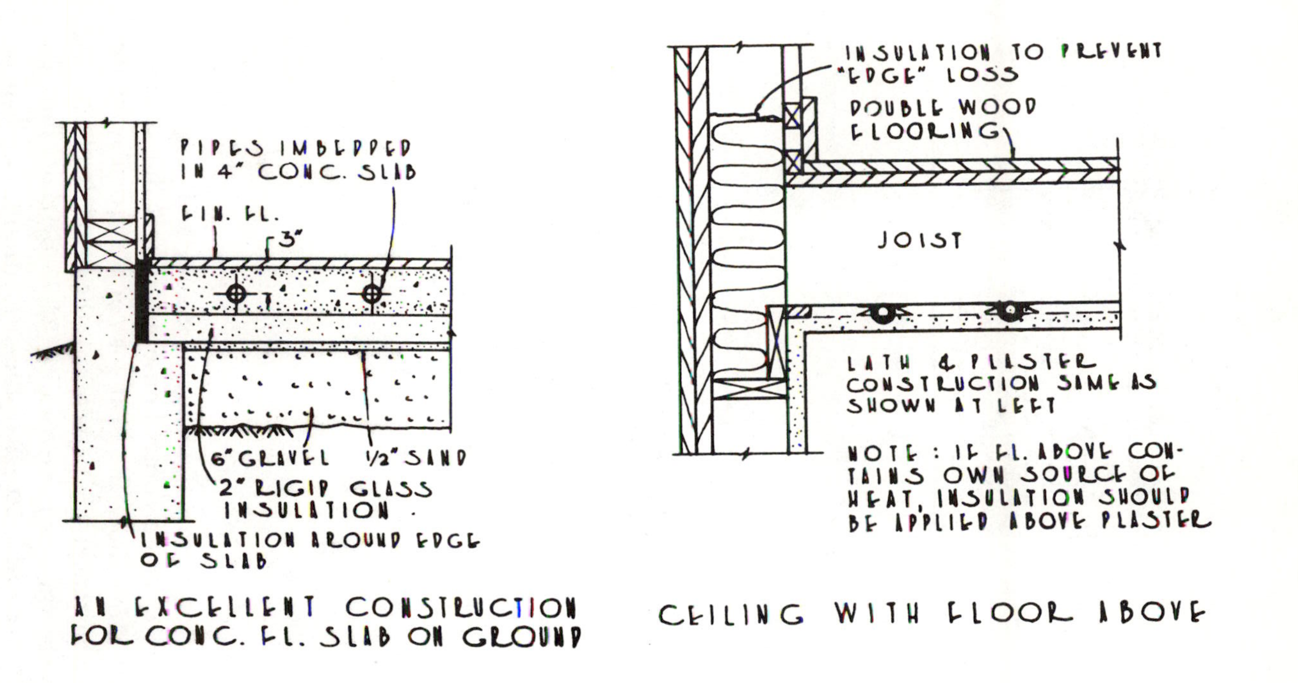

Fig. 8-17 Radiant heating piping in ceiling and floor.

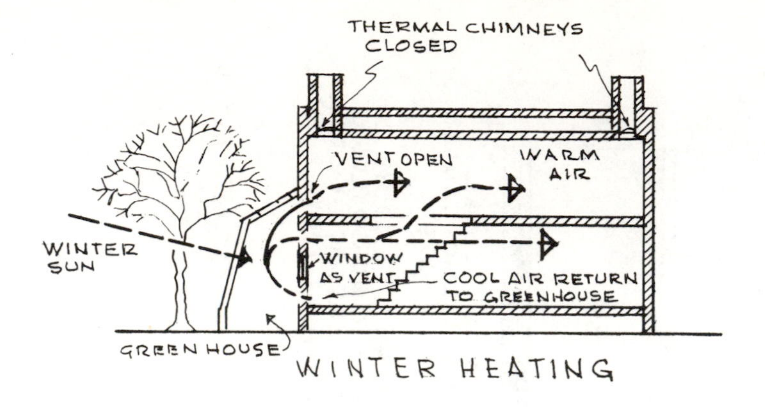

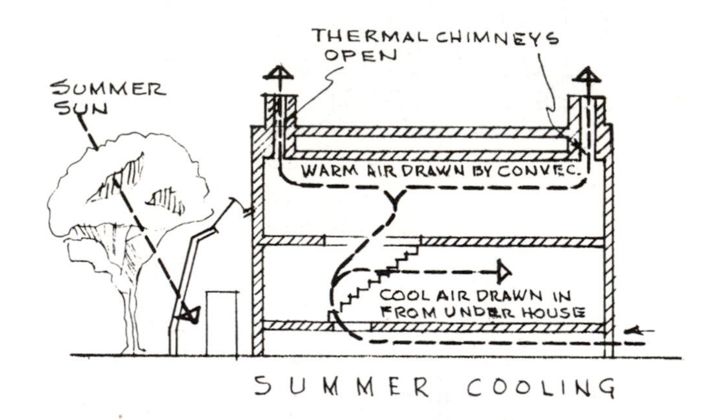

Fig. 8-18 Solar heating.

8-19 INSULATION

The homeowner of today saves considerable fuel and also has a warmer house in winter and a cooler house in summer due to insulation. Figure 8-19 shows how the heat loss of a particular house is distributed. For instance, 28 percent of the total heat loss is transmitted through the exposed walls, 21.7 percent through windows and doors, etc. Notice that the insulation alone can be effective on 42.9 percent (14.9 plus 28 percent) of the total heat loss. To save an appreciable amount of fuel, weatherstripping and storm windows should also be considered. Weatherstripping is effective on 23.9 percent of the heat loss, and storm doors and win-

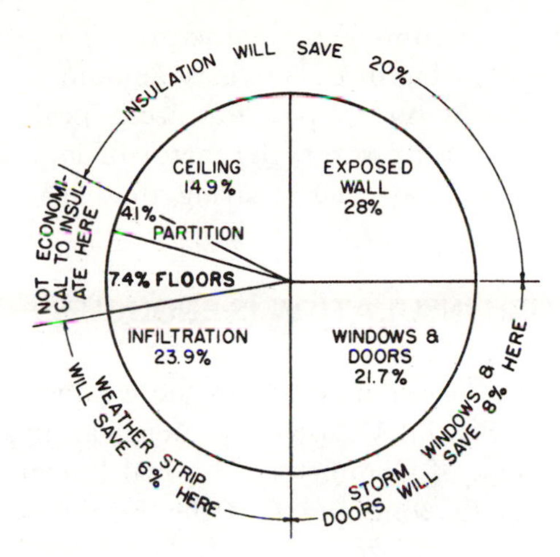

Fig. 8-19

dows are effective on 21.7 percent of the heat loss. Figure 8-20 shows the percentage saving by means of insulation and weather-stripping and use of storm doors and windows.

The total saving shown is 34 percent. Note that the insulation alone saves 20 percent of the fuel bill, but in order to capitalize on the greater savings often advertized, it is also necessary to weather-strip all doors and windows and to add storm doors and windows.

The total fuel cost saving for this particular residence would amount to about 34 percent if storm windows and doors, weather-stripping, and insulation were applied.

Fig. 8-20

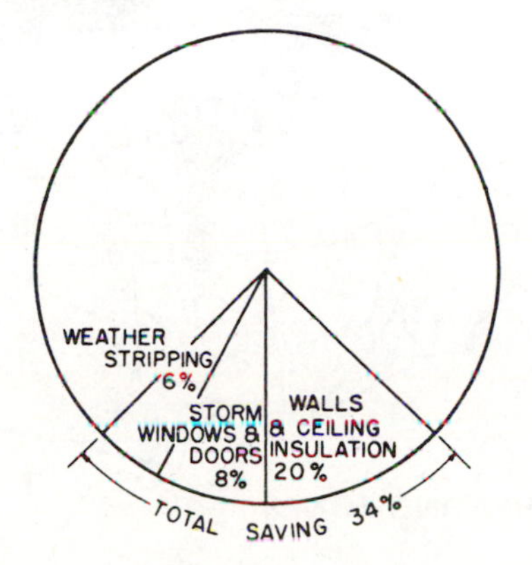

Assuming a weatherproofing job costs $2,000 and the actual fuel saving in dollars may amount to only $100 per season, it would take 20 yr to pay for itself, neglecting interest on the principal. There is however, the factor of increased comfort, and the need for present day energy saving, that will make the initial cost worthwhile.

8-20 HOW MUCH INSULATION IS ENOUGH?

The answer depends on the weather where you live. The map shown on Fig. 8-21 shows a winter heating zone map provided by the U.S. Dept. of Commerce, National Bureau of Standards. Locate your city and its zone and find the R-value in Table 8-2, for attic, wall and floor insulation.

The R-value is the thermal resistance of a building material to the flow of heat. The R-value is the reciprocal of a K- or a C-value, and represents heat flow resistance expressed in numerical values.

The K-value (thermal conductivity) represents the amount of heat which flows through one square foot of homogeneous material *one inch thick* in one hour for each degree of temperature difference between the inside and the outside temperature.

Fig. 8-21 Heating zones. (Courtesy of U.S. Department of Commerce, National Bureau of Standards.)

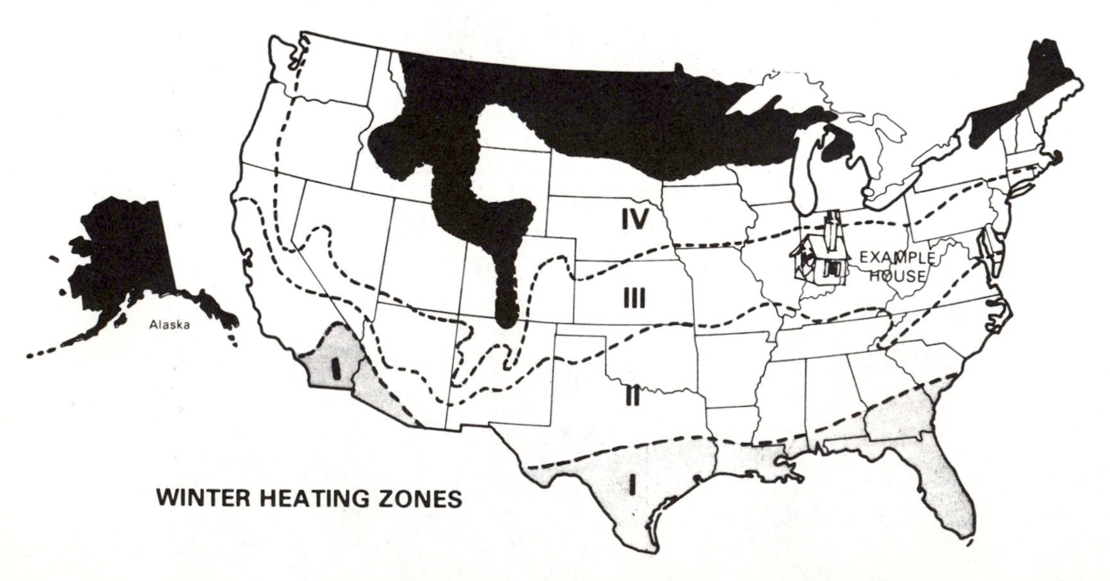

The C-value (thermal conductance) represents the amount of heat-flow through a square foot of material per hour per degree of temperature difference between the inside and the outside temperatures. The C-value is based on the stated construction or thickness of the material. This distinguishes it from the K-value, which is based on a one inch thickness of an homogeneous material having *no fixed thickness.*

The total resistance, R, of a material is used to determine its U-value, which represents the total insulating effect of a structural section, such as an insulated frame wall. The U-value is the reciprocal of the total resistance R.

EXAMPLE: In Table 8-2 for an R-22 value the U value is determined as follows:

$$U = \frac{1}{R} \quad \text{or} \quad U = \frac{1}{22} = 0.045$$

The higher the R-value, the better the insulation.

TABLE 8-2

R- and U-Values

Heating Zone	Recommended for: Ceiling		Wall		Floor	
I	R-26	U = 0.038	R-13	U = 0.076	R-11	U = 0.090
II	R-26	U = 0.038	R-19	U = 0.052	R-13	U = 0.076
III	R-30	U = 0.033	R-19	U = 0.052	R-19	U = 0.052
IV	R-33	U = 0.030	R-19	U = 0.052	R-22	U = 0.045
V	R-38	U = 0.026	R-19	U = 0.052	R-22	U = 0.045

Sources for R-value and U-value data in Table 8-2 include Federal Housing Administration; American Society of Heating, Refrigeration and Air Conditioning Engineers (ASHRAE); and Insulation Manufacturers.

8-21 ENERGY SAVING TIPS

1. The gas furnace pilot light should be turned off in spring and back on in fall. Or better yet, have an electric ignition lighter installed in the furnace as well as in the gas kitchen range.

2. The fireplace damper should be closed tightly when not in use. If there is no damper, cover the opening with a propped-up board. Caulk cracks where the fireplace meets the wall.
3. The basement door should be kept closed and weather-stripped for a tight fit. Garage doors attached to the house should be weatherstripped as well as the door leading from the garage to the house.
4. Caulk any deep masonry cracks in exterior walls.
5. Replace broken siding or shingles. To close small cracks, tack or glue pieces back in place.
6. Pipes and ducts: Check holes cut for these in floor, exterior walls, and ceiling. Seal any openings with caulk or Fiberglas insulation. (Wear gloves when handling Fiberglas.) Insulate ducts and heating pipes in unheated areas.
7. Electrical outlets and light switches on exterior walls: Cut off electricity, then remove the cover plate and pack Fiberglas insulation between the outlet box and wall.
8. Windows and doors: When they are closed, can you see cracks around the edges or feel a draft? If so, add weather-stripping. Caulk any cracks where the frame meets the wall.
9. Mail chute in door or wall: Does it close tightly? Oil its hinges.
10. Exhaust fans in bathroom and kitchen: Use as little as possible. Keep filters clean. Remove covers and caulk any major cracks around the fan.

8-22 REVIEW QUESTIONS

1. Name at least five types of heating systems.
2. State the purpose of the expansion tank.
3. Why is the two-pipe direct return hot water system not usually recommended?
4. Explain the difference between a reversed return and a direct return hot water system.

5. How many air changes per hour is the forced warm-air system capable of supplying?
6. Give the reasons for locating supply grilles on walls, 6 in. (152 mm) from the ceiling.
7. What is radiant heating?
8. What is the formula for finding the air change rate in mechanical ventilation?
9. Explain some features of solar heating.
10. What is an R-value relating to insulation?
11. List at least three energy saving methods.
12. Explain the perimeter loop duct system.

9 Equipment Selection

9-1 RADIATORS

Radiators have undergone notable changes in design, appearance, and efficiency during the last 25 years. The old fashioned cast-iron column-type radiators have given way to the modern slender tube radiator (Fig. 9-1).

The Institute of Boiler and Radiator Manufacturers (IBR) in cooperation with the Division of Simplified Practice, National Bureau of Standards, has established sizes and rating of small tube cast-iron radiators as shown in Table 9-1. The table gives standard sizes and ratings of cast-iron radiators for steam and hot water. The second column of the table shows heat output per section in square feet of equivalent direct radiation, abbreviated EDR. One square foot of equivalent radiator surface emits 240 Btu per hour when the radiator is filled with steam at 250°F and when the room temperature is 70°F. To use the table, divide the Btu per hour heat loss by 240 to arrive at the number of equivalent square feet of radiator surface required. Then select the height and width desired.

Columns A and B: Divide the catalog rating per section into the EDR required to determine number of sections.

Fig. 9-1 Modern slender-tube radiator compared with older large-tube design. (Courtesy of Burnham Boiler Corp., Irvington, N.Y.)

EXAMPLE: A certain room has a heat loss of 8000 Btu/hr. What size of steam radiator should be selected?

SOLUTION:

$$\text{Required EDR} = \frac{8000}{240} = 33.3$$

Assume a height of 25 in., which will be satisfactory with a width of 4 $\frac{7}{16}$ in. A radiator with these dimensions has a catalog rating of 2 sq ft per section.

$$\text{Number of sections} = \frac{33.3}{2} = 16.6 \text{ or } 17 \text{ sections}$$

The overall dimensions are as follows:

Length = 17 × 1¾ in. = 29 ¾ in.
(See note at bottom of Table 9-1)
Width = 4 $\frac{7}{16}$ in.
Height = 25 in.

When the radiators listed in Table 9-1 are to be used for hot water systems, the Btu per hour heat loss of the rooms is not divided by

TABLE 9-1

Standard Sizes and Ratings of Cast-Iron Radiators for Steam and Hot Water

NUMBER OF TUBES PER SECTION	CATALOG RATING PER SECTION	SECTION DIMENSIONS				
		A	WIDTH (B)		C	D
		HEIGHT	MINIMUM	MAXIMUM	SPACING	LEG HEIGHT
	SQ. FT.	INCHES	INCHES	INCHES	INCHES	INCHES
3	1.6	25	$3\frac{1}{4}$	$3\frac{1}{4}$	$1\frac{3}{4}$	$2\frac{1}{2}$
4	1.6	19	$4\frac{7}{16}$	$4\frac{13}{16}$	$1\frac{3}{4}$	$2\frac{1}{2}$
	1.8	22	$4\frac{7}{16}$	$4\frac{13}{16}$	$1\frac{3}{4}$	$2\frac{1}{2}$
	2.0	25	$4\frac{7}{16}$	$4\frac{13}{16}$	$1\frac{3}{4}$	$2\frac{1}{2}$
5	2.1	22	$5\frac{5}{8}$	$6\frac{5}{16}$	$1\frac{3}{4}$	$2\frac{1}{2}$
	2.4	25	$5\frac{5}{8}$	$6\frac{5}{16}$	$1\frac{3}{4}$	$2\frac{1}{2}$
6	1.6	14	$6\frac{13}{16}$	8	$1\frac{3}{4}$	$2\frac{1}{2}$
	2.3	19	$6\frac{13}{16}$	8	$1\frac{3}{4}$	$2\frac{1}{2}$
	3.0	25	$6\frac{13}{16}$	8	$1\frac{3}{4}$	$2\frac{1}{2}$
	3.7	32	$6\frac{13}{16}$	8	$1\frac{3}{4}$	$2\frac{1}{2}$

NOTE: OVERALL LENGTH EQUALS NUMBER OF SECTIONS TIMES $1\frac{3}{4}$ INCHES

4-TUBE SECTION SHOWN

Courtesy of the Institute of Boiler and Radiator Manufacturers (IBR).

TABLE 9-2

EDR per Sq Ft for Hot Water Radiators

Method of Firing	Average Water Temperature in Radiator	EDR per Sq Ft
Oil or gas	200°F	200

240 but by the factors shown in Table 9-2 which depend on the average water temperature in the radiator.

9-2 CONVECTORS

Convectors are radiators consisting of a nonferrous or cast-iron heating element, which is usually finned to increase the heat-transfer rate. The element is placed in a decorative metal enclosure, which may be recessed in the wall, be semirecessed or left free standing against the wall. Other units may be installed in the same place as the wood baseboard in a room and are known as baseboard heating units. Figure 9-2 shows a typical convector radiator made for use with steam or hot water. Convectors for hot water are usually rated in thousands of Btu per hour, abbreviated Mbh, whereas steam convector ratings are expressed in EDR.

Tables 9-3 and 9-4 give capacities for hot water and steam convectors, respectively.

Fig. 9-2 Trane type A convector-radiator. (Courtesy of Trane Company, La Crosse, Wis.)

TABLE 9-3

Capacities for Hot Water Convectors, Expressed in MbH (Thousands of Btu per Hour) Based on Average Water Temperature Shown

		6-in. Depth										8-in. Depth					10-in. Depth			
		Length										Length					Length			
Height	Av. Water Temp.	20″	24″	28″	32″	36″	40″	44″	48″	56″	64″	32″	40″	48″	56″	64″	40″	48″	56″	64″
	215	3.6	4.4	5.2	6.0	6.8	7.5	8.3	9.1	10.7	12.3	8.0	10.2	12.4	14.5	16.7	11.3	13.6	16.0	18.3
	210	3.4	4.2	4.9	5.7	6.5	7.2	8.0	8.7	10.2	11.7	7.6	9.8	11.8	13.9	15.9	10.8	13.0	15.3	17.5
	205	3.3	4.0	4.7	5.4	6.2	6.9	7.6	8.3	9.7	11.2	7.3	9.3	11.3	13.2	15.2	10.3	12.4	14.6	16.7
	200	3.1	3.8	4.5	5.1	5.8	6.5	7.2	7.9	9.2	10.6	6.9	8.8	10.7	12.5	14.4	9.7	11.7	13.8	15.8
20	195	2.9	3.6	4.2	4.8	5.5	6.1	6.8	7.4	8.7	10.0	6.5	8.3	10.1	11.8	13.5	9.2	11.1	13.0	14.9
	190	2.8	3.4	4.0	4.6	5.2	5.8	6.4	7.0	8.2	9.4	6.1	7.9	9.5	11.2	12.8	8.7	10.5	12.3	14.1
	185	2.6	3.2	3.7	4.3	4.9	5.4	6.0	6.6	7.7	8.8	5.8	7.4	8.9	10.4	12.0	8.1	9.8	11.5	13.2
	180	2.4	3.0	3.5	4.0	4.6	5.1	5.6	6.2	7.2	8.3	5.4	6.9	8.4	9.8	11.2	7.6	9.2	10.8	12.4
	175	2.3	2.7	3.2	3.7	4.2	4.7	5.2	5.7	6.7	7.7	5.0	6.4	7.8	9.1	10.4	7.1	8.5	10.0	11.5
	215	4.2	5.1	6.0	6.9	7.8	8.7	9.6	10.5	12.4	14.2	8.7	11.0	13.3	15.6	18.0	12.8	15.5	18.2	20.8
	210	4.0	4.8	5.7	6.6	7.4	8.3	9.1	10.0	11.8	13.6	8.3	10.5	12.7	14.9	17.2	12.3	14.8	17.4	19.9
	205	3.8	4.6	5.4	6.3	7.1	7.9	8.7	9.5	11.3	12.9	7.9	10.0	12.1	14.2	16.4	11.7	14.1	16.6	19.0
	200	3.6	4.4	5.1	5.9	6.7	7.5	8.2	9.0	10.7	12.2	7.5	9.5	11.4	13.5	15.5	11.1	13.4	15.7	18.0
24	195	3.4	4.1	4.8	5.6	6.3	7.0	7.8	8.5	10.1	11.5	7.0	9.0	10.8	12.7	14.6	10.4	12.6	14.8	16.9
	190	3.2	3.9	4.6	5.3	6.0	6.7	7.4	8.0	9.5	10.9	6.7	8.5	10.2	12.0	13.8	9.9	11.9	14.0	16.0
	185	3.0	3.4	4.3	4.9	5.6	6.2	6.9	7.5	8.9	10.2	6.2	7.9	9.6	11.3	13.0	9.2	11.2	13.1	15.0
	180	2.8	3.2	4.0	4.6	5.2	5.9	6.5	7.1	8.4	9.6	5.9	7.4	9.0	10.6	12.2	8.7	10.5	12.3	14.1
	175	2.6	3.2	3.7	4.3	4.9	5.4	6.0	6.6	7.8	8.9	5.4	6.9	8.3	9.8	11.3	8.0	9.7	11.4	13.0
	215	4.5	5.4	6.4	7.4	8.4	9.5	10.4	11.5	13.4	15.4	9.2	11.7	14.2	16.7	19.1	14.0	16.9	19.8	22.7
	210	4.3	5.2	6.1	7.1	8.1	9.0	9.9	11.0	12.8	14.7	8.8	11.2	13.6	15.9	18.3	13.3	16.1	18.9	21.7
	205	4.1	4.9	5.8	6.8	7.7	8.6	9.4	10.5	12.2	14.0	8.4	10.7	12.9	15.2	17.4	12.7	15.4	18.0	20.7
	200	3.9	4.7	5.5	6.4	7.3	8.1	8.9	9.9	11.5	13.3	8.0	10.1	12.2	14.4	16.5	12.0	14.6	17.1	19.6
32	195	3.7	4.4	5.2	6.0	6.9	7.7	8.4	9.3	10.9	12.5	7.5	9.5	11.5	13.5	15.6	11.3	13.7	16.1	18.5
	190	3.4	4.2	4.9	5.7	6.5	7.3	8.0	8.8	10.3	11.9	7.1	9.0	10.9	12.8	14.7	10.7	13.0	15.2	17.5
	185	3.2	3.9	4.6	5.3	6.1	6.8	7.5	8.3	9.6	11.1	6.6	8.4	10.2	12.0	13.8	10.0	12.2	14.3	16.4
	180	3.0	3.6	4.3	5.0	5.7	6.4	7.0	7.8	9.0	10.4	6.2	7.9	9.6	11.2	12.9	9.4	11.4	13.4	15.4
	175	2.8	3.4	4.0	4.7	5.3	5.9	6.5	7.2	8.4	9.7	5.8	7.3	8.9	10.4	12.0	8.7	10.6	12.4	14.2

Courtesy of Trane Company, La Crosse, Wis.

TABLE 9-4

Capacities for Steam Convectors, Sq Ft EDR
With 215°F Steam and 65°F Entering Temperature of Air

	6-in. Depth										8-in. Depth					10-in. Depth			
	Length										Length					Length			
Height	20″	24″	28″	32″	36″	40″	44″	48″	56″	64″	32″	40″	48″	56″	64″	40″	48″	56″	64″
20	16.0	19.5	23.0	26.5	30.0	33.4	37.0	40.5	47.5	54.5	35.5	45.5	55.0	64.5	74.0	50.0	60.5	71.0	81.5
24	18.5	22.5	26.5	30.5	34.5	38.5	42.5	46.5	55.0	63.0	38.5	49.0	59.0	69.5	80.0	57.0	69.0	81.0	92.5
32	20.0	24.0	28.5	33.0	37.5	42.0	46.0	51.0	59.5	68.5	41.0	52.0	63.0	74.0	85.0	62.0	75.0	88.0	101.0

Courtesy of Trane Company, La Crosse, Wis.

EXAMPLE: A certain room has a heat loss of 9150 Btu per hour. The hot water boiler is oil-fired. The height of convector is not to be over 22 in. and not deeper than 6 in. Select a convector of the free-standing type.

SOLUTION: Probable average water temperature in radiator (Table 9-2) = 200° F

$$\text{Mbh output} = \frac{9150}{1000} = 9.15$$

Referring to Table 9-3, capacities for hot-water convectors, for 6-in. depth and 20-in. height, find 9.2 Mbh for 56-in. length.

Overall dimensions (see Fig. 9-3)

Height = 20 in.
Length = 56 in. + ⅜ in. = 56 ⅜ in.
Depth = 6 in.

Note that the actual cabinet lengths for free standing steam and hot-water convectors are always ⅜-in. (9.5 mm) longer than reference length. Heights are exactly 20 in. (508 mm), 24 in. (610 mm), or 32 in. (813 mm). Recesses should be constructed ½-in. (12.7 mm) longer than overall length and ¼-in. (6.4 mm) higher than height of

Fig. 9-3 Dimensions for freestanding steam and hot water convectors.

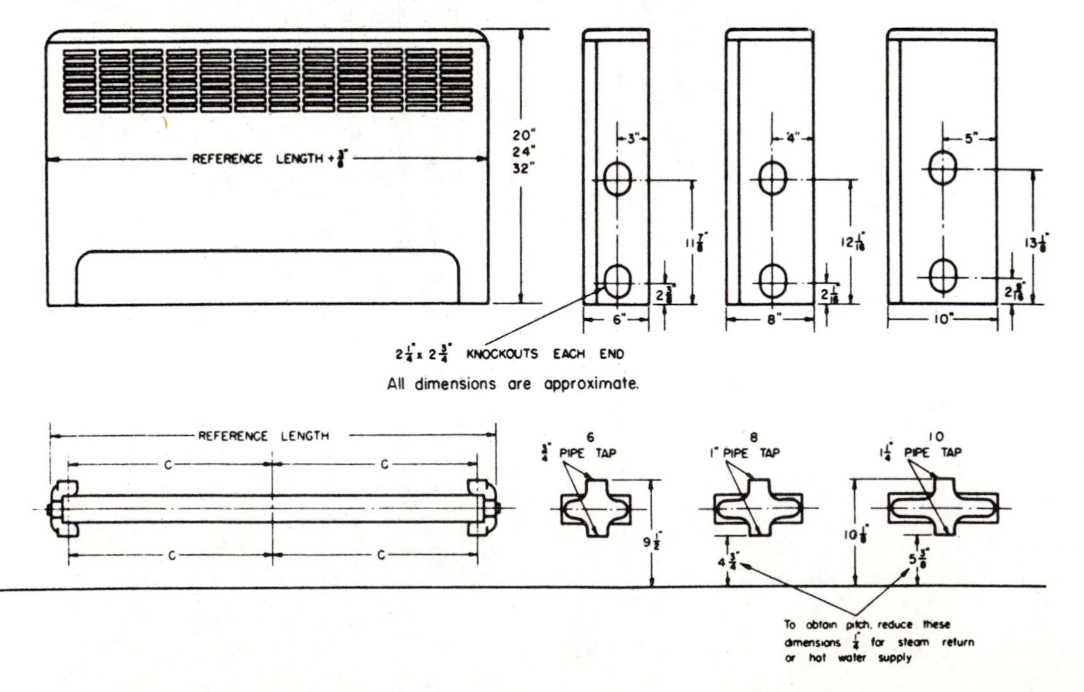

the unit. If a convector radiator is recessed into an exposed wall, the back of the unit should be covered with ½-in. insulation.

The two knockouts on each end of the cabinet are located equidistant from the front and back of the cabinet. Cabinets can be recessed any distance up to 1¼ in. (31.8 mm) from the front. This permits easy removal of the front panel.

Side connections should be avoided on recessed installations if a unit with a 6-in. (152 mm) depth is used.

Steam and Hot Water Convectors

Typical piping connections for steam convectors on two-pipe system are shown in Fig. 9-4. If it is desired to use a convector on a one-pipe system, a special air vent is required and the return trap is omitted. Connections for a hot-water convector are shown in Fig. 9-5.

Baseboard heating units (Fig. 9-6) are installed in place of the wood baseboard in a room. They are made of hollow cast-iron placed along the outside walls of a room. This type of heating unit has metal fins, closely spaced and fastened around the hot water pipe. Baseboard units such as those shown in Fig. 9-7 distribute heat evenly through the room.

Electric heating is not much different from other systems. It uses baseboard heaters and convectors like those in hot water systems. Instead of circulating water, heating elements such as those on electric kitchen ranges are used.

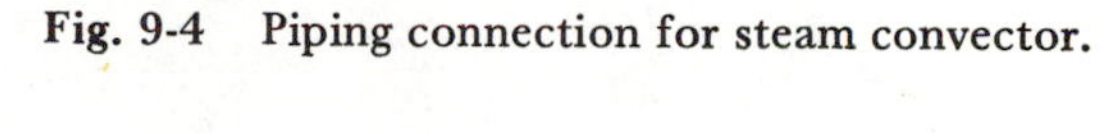

Fig. 9-4 Piping connection for steam convector.

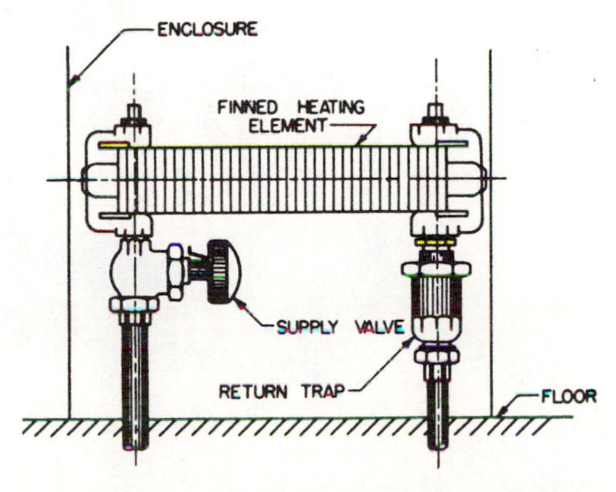

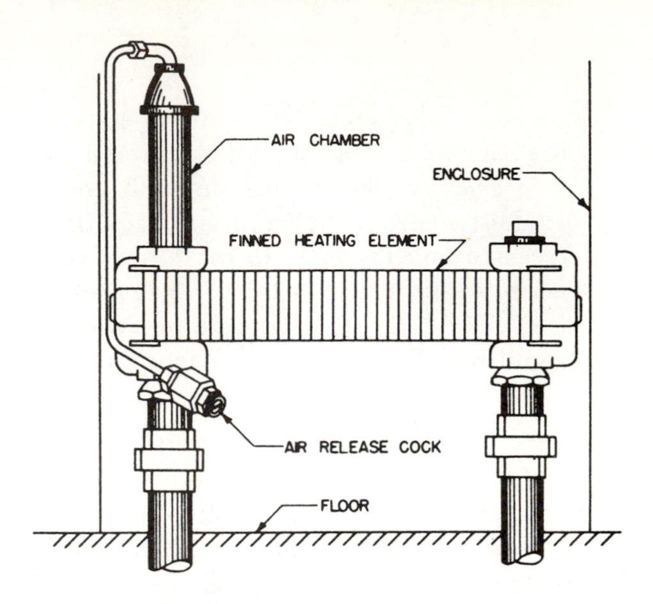

Fig. 9-5 Piping connection for hot water convector.

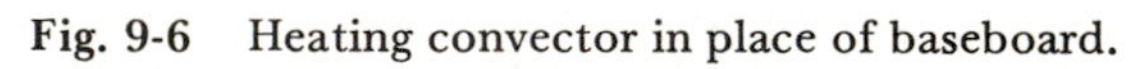

Fig. 9-6 Heating convector in place of baseboard.

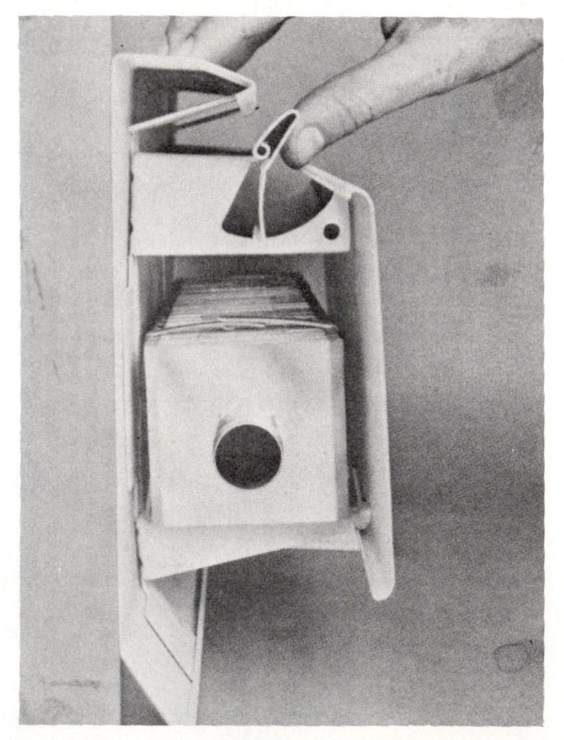

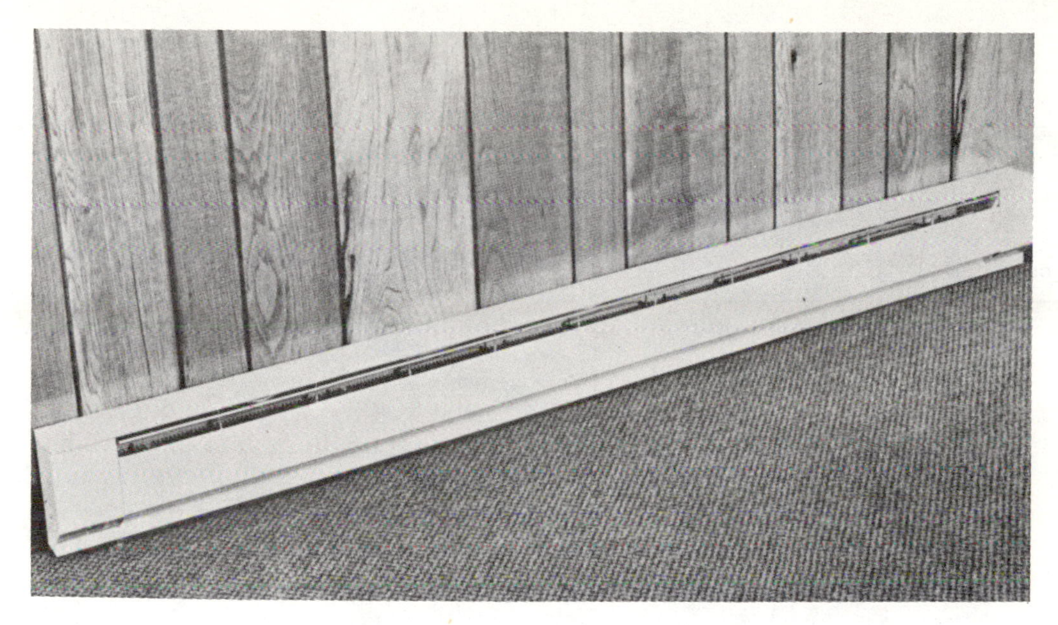

Fig. 9-7 Baseboard heating unit.

The electric elements in baseboard heaters run the length of the heater as shown in Fig. 9-8. There are no glowing coils because the elements are embedded in cast aluminum, and there are no fins to collect dust or lint. An electric terminal box at one end allows for the electric connection.

Baseboard electric heaters come in many different lengths and are made to fit snugly against the wall and floor. They take the place of the room baseboard. As in other heating systems, the Btu heat loss must first be found before a convector length can be found.

9-3 FORCED WARM-AIR GAS FURNACES

The furnace (Fig. 9-9) contains no water and is used to heat air for a warm-air system as well as provide the heat for the domestic hot water tank usually adjacent to the furnace. Domestic hot water may also be heated by a separate gas or electric automatic water heater.

Between the furnace and the supply grilles there is a heat loss through the sheet metal ducts carrying the warm air, and consequently more Btu's are needed at the furnace bonnet than are required by the hourly heat loss. Further, since the fresh air brought in from

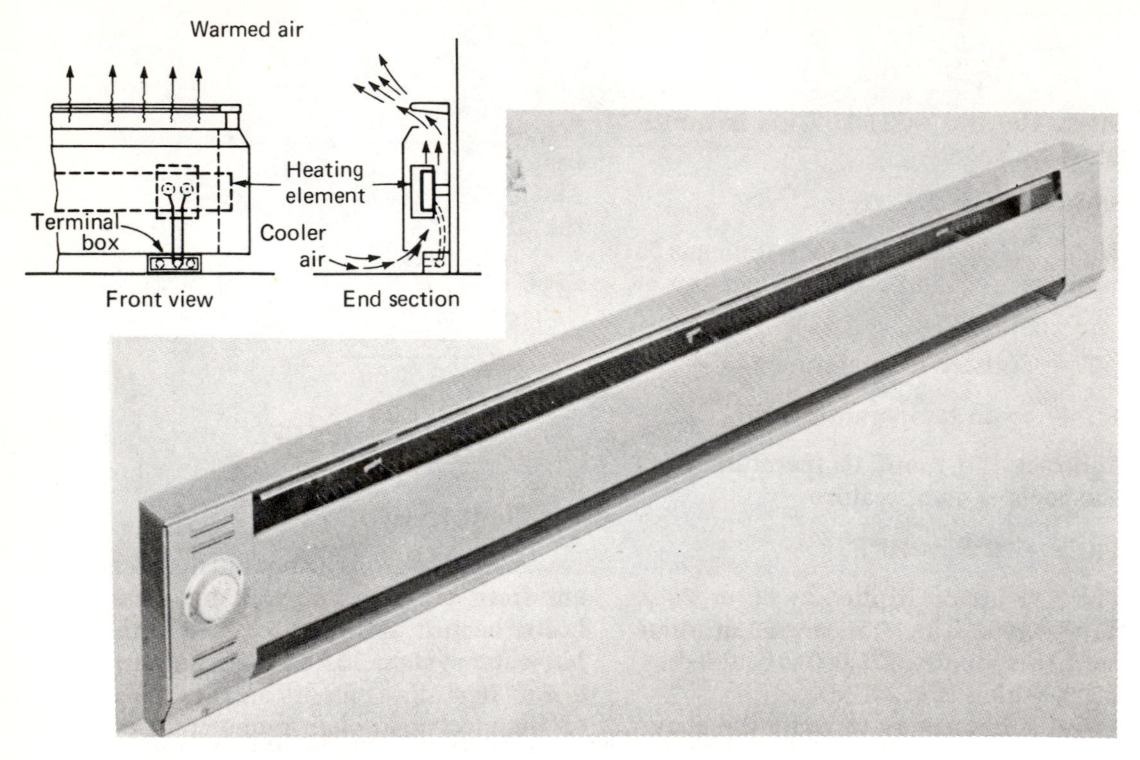

Fig. 9-8 Electric baseboard heating unit.

outside must be heated, an additional load on the furnace is imposed. Manufacturers show a Btu delivery at the bonnet which is in excess of the net Btu by an amount that has been allowed for the duct heat loss plus the fresh air load. Here again, when this allowance is not sufficient, because of abnormal amounts of fresh air or excessively long duct runs, the manufacturer should be consulted.

9-4 DUCT OUTLETS

Figure 9-10 shows a floor diffuser with a lever-control rod and blades. The amount of warm air can be regulated or shut off entirely. Floor outlets of this type may interfere with furniture or wall to wall carpeting. Figure 9-11 shows a round duct and a transition duct piece to which a floor diffuser is fixed.

Fig. 9-9 Forced warm-air gas furnace.

9-5 AUTOMATIC CONTROL: NIGHT SET-BACK THERMOSTAT

Figure 9-12 shows a night set-back thermostat that automatically sets a lower temperature during the night, and automatically sets a higher temperature in the morning. The following are the major parts:

A solid state quartz crystal timer (1) that operates by setting

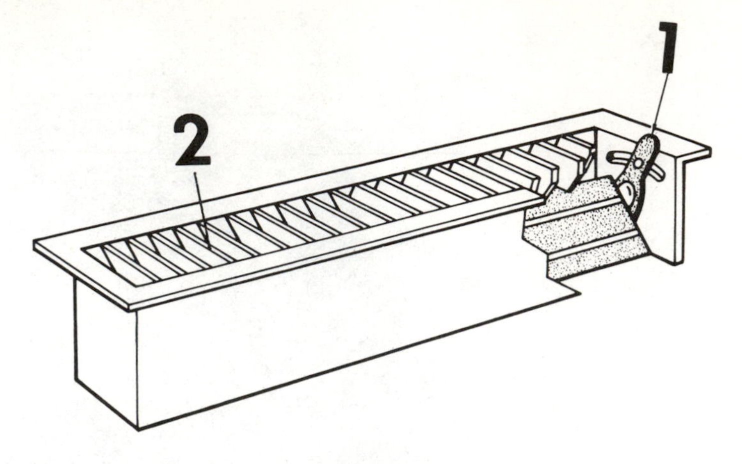

1. Lever control rod
2. Blades

Fig. 9-10 Floor diffuser.

Fig. 9-11 Floor location for warm-air outlet.

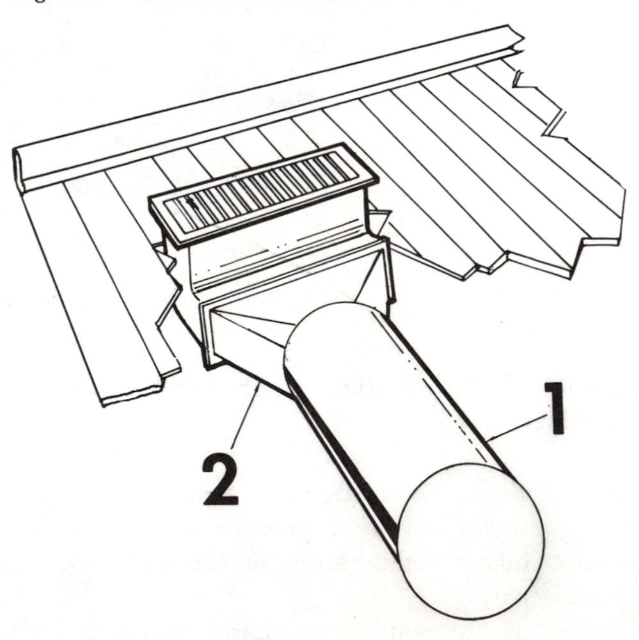

1. Duct
2. Transition piece

Fig. 9-12 The night set-back thermostat. (Courtesy of White-Rodgers Division, Emerson Electric Company.)

the pointer on clock to morning or night time that you want the lower temperature to come into effect.

The other pointer (2) is set to the time you want higher temperatures to resume.

The system and fan controls are at (3). Switch levers are easy to see and positively "snap" into the desired position.

The high temperature selection lever is (4). Set this lever to temperature you desire for the high temperature level. When the quartz clock reads the time you have selected with the clock pointer (1), this temperature automatically goes into effect.

The low temperature selection lever is (5). Set this lever at the lower temperature level you desire. When clock reaches time you have selected with the clock pointer (2), this lower temperature automatically goes into effect. The thermostat will continue to raise and lower the temperature to these settings automatically until you change the settings. A level bubble is provided in the wall plate for precise mounting.

The sensing elements and switches are at (6). Proven sensitive spiral bimetal sensing elements and sealed mercury switches are used. A nickel cadmium automatically recharged battery powered timer is maintained at full power automatically by the normal thermostat circuit.

9-6 REVIEW QUESTIONS

1. What does IBR represent?
2. What does EDR represent?
3. A certain room has a heat loss of 12,000 Btu/hr. What size steam radiator should be selected? Assume a height of 25 in. and a width of $4\frac{7}{16}$ in.
4. What is done to convectors to increase the heat transfer rate?
5. How are hot water capacities expressed?
6. Baseboard heating units are made of what material?
7. Heating elements are used in what type of baseboard heater?
8. Name two types of duct outlets.
9. Briefly explain the night set-back thermostat.
10. What is provided in the night set-back thermostat to allow precise mounting?

10 Plumbing and Heating Review Questions, Answers, and Problems

10-1 QUESTIONS

1. Define each of the following terms as used by the plumbing trades and explain the purpose of each.
   (a) Catch basin
   (b) Cleanout
   (c) Backwater prevention valve
   (d) Hydraulic grade
   (e) Ferrule plate
2. As used by the plumbing trade, what is meant by each of the following?
   (a) Soil pipe
   (b) Waste pipe
   (c) House sewer
   (d) House drain
   (e) Vent
3. Answer both (a) and (b).
   (a) What is the function of a trap in a plumbing system?
   (b) What causes siphonage of traps and how is it prevented?
4. What are the essential processes involved in a sewage disposal system?

5. Define briefly and illustrate by a simple sketch each of the following:
   (a) Wet vent
   (b) Fire damper (as applied to duct work)
   (c) Expansion joint (as applied to piping)
   (d) Aquastat
6. If a floor drain has to be connected to a sanitary sewer line, what added provisions must be made? State the reasons.
7. (a) How will the available water pressure at street level offset the design of a building's water supply system?
   (b) What are the commonly accepted methods of testing (1) soil, (2) water piping, and (3) gas lines in new construction? When should these tests be made?
8. Name eight kinds of pipe, manufactured of different materials, and describe the use that you would recommend for each.
9. Why are vents installed on drains from plumbing fixtures? Why is a fresh air inlet used in a plumbing system?
10. List all the piping of both the (a) water supply and (b) drainage systems of a house from the sewer to the roof. State the material of which each is made and explain why that material was used.
11. In a suburban development what would you recommend as the best solution to problem involved in sewage disposal where the public sewer main is 8 ft above the level where the house sewer line leaves the residence?
12. State what toilet facilities (with a list of fixtures) you would provide for a 10-story office building having 10,000 square feet per floor gross area, with 25 women and 25 men per floor.

10-2 ANSWERS

1. (a) A catch basin or drainage inlet is built below a paved surface. It may be of brick, concrete block, concrete, or precast concrete. A screened opening at the top permits the rain water to enter the catch basin from which it flows by drain pipe to the storm sewer, combination storm, and sanitary sewer, or to underground basins that disperse the water back into the ground. The top opening of the catch basin may be

2 sq ft (0.186 m^2) and 3 sq ft (0.279 m^2) at the bottom with a minimum depth of 3 ft (914 mm).

(b) In order to prevent clogging in the house drain, cleanouts must be provided at intervals not more than 50 ft (15.24 m) apart and at the end of the house drain beyond the last vertical stack, and at each change of direction of horizontal run.

(c) A back water prevention valve (BWPV) is provided to prevent storm water from backing up into a storm system, into areaway drains, or yard drains. When street storm drains overflow due to heavy rains the water tends to push into the storm drain when it is lower than the street. The BWPV is designed to shut the valve when the flow is opposite its normal flow.

(d) Hydraulic grade means a line connected from the high water level of a fixture, such as a sink to the branch connection of the soil stacks. Refer to Fig. 2-7.

(e) The ferrule plate is fastened to the lead bend, which receives the ferrule gasket before the water closet is placed. The water closet is bolted to the ferrule gasket and ferrule. Refer to Fig. 6-7.

2. (a) The soil pipe is one into which drains soil or solid waste matter.

(b) The waste pipe receives only waste water from sinks, lavatories, and tubs.

(c) The house sewer is the sewer pipe that extends from 2 to 3 ft (610 to 914 mm) from the house to the public sewer in the street.

(d) The house drain is the horizontal sewer pipe within the house that receives the soil and waste stacks.

(e) Vents are pipes that lead from fixture traps and through the roof.

3. (a) To prevent the passage of odors and vermin.

(b) Siphonage of a fixture trap can best be explained by referring to Fig. 2-8. A tube is used for carrying liquid out over the top edge of a container. This occurs through the force of atmospheric pressure upon the surface of the liquid. One end of the tube is placed in the liquid and the other end in a container placed at a lower level. The tube must be filled by suction before flow will start. If valve A is opened, the siphoning action will stop. Similarly, a vent at B will stop the siphoning of the trap when atmospheric pressures are equalized.

4. A sewage disposal system must have a septic tank into which the house sewage flows. Its size is designed based on the occupancy of the house. The solid matter settles to the bottom of the tank, while the effluent is led into a distribution box and then into leaching pool consisting of an open-jointed serpentine system of pipe that distributes the effluent into the soil. The piping length of the leaching system is determined by the occupancy using the system.

5. (a) A wet vent is that portion of a vent pipe through which liquid waste water flows.
   (b) Fire dampers are placed inside a duct system. The damper has a fusible link, which melts and closes the damper when approached by fire or great heat. It is usually placed between partition walls to prevent the spread of fire.
   (c) An expansion joint is a pressure-tight device which permits expansion or contraction of pipe lines.
   (d) An aquastat is a hot water control, such as insertion or surface thermostats, designed for application to hot water heating systems.

6. The floor drain must have a trap, preferrably a deep seal trap. Also, a water valve should be within approximately 10 ft (3 m) to refill the trap.

7. (a) Water pressures at street level, say 40 psi (275.8 kPa) will limit the height of a building, unless the water is collected in a tank in a cellar or basement and then pumped up to a water tank on the roof of the building.
   (b) (1) Soil: Visual inspection
   Loading platform
   Test borings

   (2) Water piping: The system can be tested under a water pressure not less than the working pressure under which it is to be used. A 50 psi (344.5 kPa) air pressure may be substituted for the water test.

In either method of testing, the piping shall withstand the test without leaking for a period of not less than 15 min. The test should be made just before the system is to be used.

(3) Gas Piping: This is to be an air pressure test where the piping shall stand a pressure of not less than 10 psi (68.9 kPa) gauge pressure, or a pressure of at least six in. (152 mm) of mercury, measured with a manometer or slope gauge. The test pressure shall be held in no case less than 15 min, with no perceptible drop in pressure. The test should be made when the system is ready for use.

8. PVC (polyvinyl chloride) pipe has a broad range of applications, including water mains and water service, sewer mains and building sewers, gas distribution and service, irrigation, electrical and communication conduit, and industrial process piping.

PE (polyethylene) pipe is used for water service, gas distribution and service, irrigation, and chemical waste systems.

CPVC (chlorinated polyvinyl chloride) is used for domestic hot and cold water piping systems as well as for industrial process piping applications.

DWV (drain, waste, vent) plastic sanitary drainage piping has found great acceptance with residential construction in the United States. It is gaining rapid acceptance in commercial, industrial, and institutional buildings as well. Plastic piping can also be used for industrial process piping, irrigation, and sprinkler systems.

Galvanized steel pipe has been used on older water supply systems. It is good for withstanding the pressures of water hammer and high water pressure. It rarely leaks because the threaded connections tend to seal each other.

Cast-iron pipe is used for drain waste and vent piping, and is preferred by most plumbers. Cast iron is strong and has great resistance to corrosion.

Brass pipe is more expensive than other pipe. It is mostly used for water supply where the water is particularly corrosive.

Copper pipe is furnished as plumbing tubing, and includes types K, L, M, and DWV. Type L pipe is ACR tubing, for air conditioning and refrigeration service.

9. To prevent the siphoning of traps.
   Fresh air inlets prevent the siphoning of the house trap.

10. (a) Water Supply Piping:
    (i) Service line
    (ii) House main
    (iii) House branches
    (iv) Risers
    (v) Fixture branches

    The above water lines, including fittings, should be of brass, copper, cast iron, galvanized malleable iron, galvanized wrought iron, galvanized steel, lead, or other approved materials. PE or PVC water pipe manufactured to recognized standards may be used for cold water distribution systems outside of the building. The materials selected should be compatible with the type of water in the area. Always check with local administrative authorities.

    (b) Drainage System:
    (i) House sewer: cast iron (B&S) (Bell & Spigot), plastic
    (ii) House trap and house trap vent stack: cast iron (B&S), plastic
    (iii) House drain: cast iron (B&S), plastic
    (iv) Soil stack: cast iron (B&S), plastic
    (v) Waste stack: cast iron (B&S), plastic
    (vi) Fixture branches: cast iron (screwed), plastic
    (vii) Fixture traps: plastic, copper, cast iron
    (vii) Fixture vents: plastic, copper, cast iron (screwed)
    (ix) Vent stack through roof: cast iron (B&S)

11. Provide for a sump pump. The sewage flows into a pit from which it is lifted or pumped to a level for an ordinary gravity flow to the public sewer.

12. Male: 20 w.c., 20 lav.
    Female: 20 w.c., 20 lav.
    According to the Uniform Plumbing Code (IAPMO) 1979.

10-3 TESTING THE PLUMBING SYSTEM

There are two methods of testing a new plumbing system: the water test, and the air test.

Water Test

This test is applied to the drainage and vent systems either in its entirety or in sections. If applied to the entire system, all openings in the piping must be tightly closed except the highest opening, and the system filled with water to point of overflow.

If the system is tested in sections, each opening shall be tightly plugged except the highest opening of the section under test, and each section shall be filled with water, but no section should be tested with less than 10-ft head of water. The water shall be kept in the system, or in the portion under test, for at least 15 min before the inspection starts, and the system must then be found tight at all points.

Air Test

This test is made by attaching an air compressor testing apparatus to any suitable opening, and, after closing all other inlets and outlets to the system, forcing air into the system until there is a uniform gauge pressure of 5 psi (34.5 kPa) or sufficient to balance a column of mercury 10 in. (254 mm) in height. The pressure must be held without introduction of additional air for a period of at least 15 min.

10-4 PLUMBING RESEARCH PROBLEMS

P. 1. Indicate by diagrammatic sketches the complete plumbing system in a one-story and basement residence in the country where there are no sewers and where there is no water supply from a local utility company. Show all waste, vent, and sewer lines and indicate the sewage disposal system and water supply piping. Indicate materials for all piping. Assume the following:

One bathroom with one w.c., one lav., and one bathtub. One kitchen sink plus one small w.c. and lav. room off kitchen. One floor drain in basement.

Freehand isometric sketches or sketch plans may be used to illustrate the solution.

P. 2. A boy's school in the country is building a new two-story and basement dormitory for 40 students. All student washing and toilet facilities will be grouped into one area on each floor.

(a) Draw a schematic diagram of both floors with the required number of water closets, lavatories, stall and gang showers and their waste, soil, and vent pipe connections. Disregard water piping. Give the size and material for all piping.

(b) Draw a diagram of a sewage disposal system to handle the building. What are the factors that determine the size of the units of the disposal system?

P. 3. Sketch all the piping and rough-in required for a bathroom in a private home, starting at the sewer and water mains in the street and continuing to the top of the vent stack and to each fixture. Note the materials that should be used.

P. 4. An apartment house will have its top floor 108 ft (33 m) above the level of the water mains in the street. Explain the water supply system you would use assuming 40 psi (275.8 kPa) of water pressure at the main.

P. 5. Sketch a complete sanitary disposal system (outside building) for a small industrial plant employing 75 persons. This plant has ample acreage for all facilities, the soil is heavy loam, and the site is nearly level.

P. 6. Describe the difference between a wet and dry sprinkler system and state where you would use each system.

P. 7. What is a standpipe system and how does it function?

P. 8. How would you take care of sewage disposal and water supply for a six-classroom rural school, where no municipal services for these are available? Describe briefly the principal parts of each part.

P. 9. What materials, size, and method of joining would you choose for the following piping in an eight-room residence?

(a) Hot water lines
(b) Cold water lines
(c) House sewer
(d) Vent stacks
(e) Gutters and leaders

P. 10. Describe how you would ventilate a moving-picture theater that seats 600 persons.

Heating

P. 11. Describe briefly what is meant by the term "zones" in a heating system and give an example of how they are used.

P. 12. From the following types of heating systems, select the one you would use in each of the buildings listed below. Give a reason for your choice.

Systems: Forced hot water circulating pump
Steam (vacuum)
Radiant hot water
Forced warm-air
One-pipe hot water

Buildings: Elementary school (16 classrooms)
Four-family city apartment building
Candy shop
Small theater
Country residence

P. 13. Make a diagram of a radiant heating system that uses coils in the floor. The floor is at ground level. Illustrate the design of the floor slab.

P. 14. Sketch an oil burning boiler as required for a forced recirculating hot water heating system. Indicate boiler connections and principal safety features.

P. 15. Name several kinds of buildings that require fixed or constant air temperatures and humidity.

Appendix

ABBREVIATIONS

ABS: Acrylonitrile butadiene styrene
AGA: American Gas Association
AHAM: Association of Home Appliance Manufacturers
AISI: American Iron and Steel Institute
ANSI: American National Standards Institute
API: American Petroleum Institute
ASA: American Standards Association, now ANSI
ASME: American Society of Mechanical Engineers
ASTM: American Society for Testing and Materials
AWWA: American Water Works Association
CISPI: Cast-Iron Soil Pipe Institute
CS Commodity Standards Division of U.S. Department of Commerce
CS&PS: Commercial Standards and Product Standards
Fed: Federal Specifications, U.S. Government Department of Commerce

IAPMO: International Association of Plumbing and Mechanical Officials

IBR: Institute of Boiler and Radiator Manufacturers

EDR: Effective direct radiation

MIL: Military Specifications, U.S. Government

MSS: Manufacturers Standardization Society of the Valve and Fittings Industry

NSF: National Sanitation Foundation

JAN: Joint Army-Navy Specifications, U.S. Government, Department of Defense

PB: Polybutylene

PDI: Plumbing and Drainage Institute

PS: Material and Property Standard published by IAPMO

PVC: Polyvinyl chloride

SBI: Steel Boiler Institute

SWP: Solvent welded pipe

UL: Underwriters' Laboratories

UPC: Uniform Plumbing Code published by IAPMO

WQA: Water Quality Association

ID: Inside diameter

OD: Outside diameter

HTN: Heat transmission number

METRIC SYSTEM

International System of Units (SI)

Conversion Table

To Convert	Into	Multiply by
Btu	joules	1.055
Btu/hr	watts	0.2931
Btu/min	kilowatts	0.01757
Btu/min	watts	17.57
circumference	radians	6.283

Conversion Table (Continued)

To Convert	Into	Multiply by
cubic feet	cubic metres	0.02832
cubic feet	litres	28.32
cubic centimetres	cubic inches	0.06102
cubic feet/min	cu cm/sec	472.0
cubic inches	cubic cms	16.39
cubic metres	gallons (US)	264.2
degrees Celsius (°C)	degrees Fahrenheit (°F)	$(C - 32) \times 9/5$
feet	centimetres	30.48
feet	metres	0.3048
feet	millimetres	304.8
gallons	litres	3.785
horsepower	watts	745.7
horsepower-hr	kilowatt-hr	0.7457
kilograms	pounds	2.205
kilometres	miles	0.6214
kilometres/hr	miles/hr	0.6214
kilowatt-hrs	Btu	3.413
litres	cubic feet	0.3531
litres	gallons (U.S.)	0.2142
metres	feet	3.281
metres	inches	39.37
metres	yards	1.094
ounces (fluid)	litres	0.02957
pounds	kilograms	0.4536
psi	kilopascals	6.895
quarts (liquid)	litres	0.9463
radians	degrees	57.30
square inches	square millimetres	645.2
watts	Btu/hr	3.4129
watts	Btu/min	0.05688

DEFINITIONS

Angle Valve: Similar to a globe valve but pipe connections are at right angles.

Areaway: A wall surrounding a window below grade allowing light to enter.

Aquastat: An automatic switching device consisting of a metal- or liquid-filled heat-sensitive element designed to detect temperature drop or rise of the boiler water.

Bushing: A tapped fitting which is used to reduce the size of an end opening of a fitting or a valve.

Catch Basin: A cast-iron, concrete, or wooden receptacle into which the water from a roof, floor, etc., will drain. It is connected with a sewer or drain tile.

Check Valve: A valve that closes automatically when the flow in a pipe is reversed.

Conductor Pipe: A round, square, or rectangular metal pipe used to lead water from the roof to the sewer.

Convector: A heat-transfer surface designed to transfer its heat to surrounding air largely by convection currents.

Corrosion: The deterioration of piping materials due to a chemical action or galvanic action.

Drain: A means of carrying off waste water. A sewer or other pipe used for conveying ground, surface, or storm water or sewage.

Drywell: A pit located on porous ground walled up with rocks which allows water to seep through the pit. Used for the disposal of rain water or the effluent from a septic tank.

Elbow: A pipe fitting made to allow a turn in direction of a pipe line.

Expansion Bends: A loop in a pipe line that permits the expansion and contraction of the pipe.

Female Thread: Internal threads.

Fixture: A receptacle attached to a plumbing system in which water or other waste may be collected for ultimate discharge into the plumbing system.

Flange: A rimlike end on a valve, or pipe fitting for bolting another flanged fitting. Usually for large diameter on pressurized pipe.

Gate Valve: A valve which regulates flow within a pipe. It is known as an on-off valve.

Globe Valve: Allows for throttling the flow of water.

Gutter: A trough for carrying off water.

Hanger: Supports specially designed to support pipe lines.

House Drain: That part of the horizontal sewer piping inside the building receiving waste from the soil stacks.

Hub-End: Pipe end connections that are leaded and caulked such as on cast-iron sewer piping.

Increaser: A short pipe fitting with one end of a larger diameter.

Joint: Where two pipes are connected either by bolting, welding, or by a screwed connection.

Lavatory: A basin for washing hands and face.

Louvers: A series of shutters used for the circulation of air.

Male Thread: An external pipe thread.

Manhole: An opening constructed in a sewer to allow access for a person.

Manometer: An instrument (usually a V-shaped tube) used to measure pressure of gases, liquids, and vapors.

Nipple: A short length of pipe threaded at both ends to allow for joining pipe elements.

Nonrising Stem Valve: A valve where the stem does not rise when opening the valve.

Packing: Material used in stuffing box of a valve to keep a leak-proof seal around the stem.

Plug: A cap used for shutting off a tapped opening.

Pressure Regulator: A valve used to automatically reduce and maintain pressure.

Reducer: A pipe fitting with a smaller opening at one end.

Relief Valve: A valve that will automatically open when the pressure inside a vessel or container exceeds a specified amount.

Rising Stem Valve: A valve whose stem rises when the valve is opened.

Septic Tank: A concrete tank, embedded in the earth into which sewage is allowed to drain.

Soft Jaws: Covers of lead or copper placed over vise jaws to prevent damage to materials held in a vise.

Solder Joint: A connection of piping made by soldering. Generally used with copper tubing.

Soil Stack: The vertical pipe in the house plumbing system into which sewage from fixtures and branches discharges.

Tap: A tool for forming internal or female threads.

Tee: A three-way fitting shaped like the letter T.

Temperature: Heat and cold recorded in degrees on a thermometer.

Thermostat: An automatic device for controlling the supply of heat.

Threader: A device or tool used to cut threads on the end of a piece of pipe.

Trap: A U-shaped pipe filled with water and located beneath plumbing fixtures to form a seal against the passage of foul odors or gases.

Tubing: Lightweight pipe made of materials such as copper, brass, or plastic.

Union: A type of fitting used to join lengths of pipe for easy opening of a pipe line.

Valve: A device designed to regulate the flow of fluids or gases.

Vent Pipe: A vertical pipe used to ventilate plumbing systems and to provide a release for pressure caused by flushing.

Vent Stack: The upper portion of a waste or soil stack above the highest fixture.

Vitreous: A material that resembles glass, such as lavatories or drain pipe.

Waste Stack: A plumbing pipe used to receive liquid discharge from lavatories, sinks, and bathtubs.

Converting Litres to U.S. Gallons

l	US gal	*l*	US gal	*l*	US gal	*l*	US gal
1	0.284	26	6.869	51	13.473	76	20.077
2	0.528	27	7.133	52	13.737	77	20.341
3	0.793	28	7.397	53	14.001	78	20.605
4	1.057	29	7.661	54	14.265	79	20.870
5	1.321	30	7.925	55	14.530	80	21.134
6	1.585	31	8.189	56	14.794	81	21.398
7	1.849	32	8.454	57	15.058	82	21.662

Converting Litres to U.S. Gallons (Continued)

l	US gal	l	US gal	l	US gal	l	US gal
8	2.113	33	8.718	58	15.322	83	21.927
9	2.378	34	8.982	59	15.586	84	22.190
10	2.642	35	9.246	60	15.850	85	22.455
11	2.907	36	9.510	61	16.115	86	22.719
12	3.170	37	9.774	62	16.379	87	22.963
13	3.434	38	10.039	63	16.643	88	23.247
14	3.698	39	10.303	64	16.907	89	23.511
15	3.963	40	10.567	65	17.171	90	23.776
16	4.227	41	10.831	66	17.435	91	24.040
17	4.491	42	11.095	67	17.700	92	24.304
18	4.755	43	11.359	68	17.964	93	24.569
19	5.019	44	11.624	69	18.228	94	24.832
20	5.283	45	11.888	70	18.492	95	25.097
21	5.548	46	12.152	71	18.756	96	25.361
22	5.812	47	12.416	72	19.020	97	25.625
23	6.076	48	12.680	73	19.285	98	25.889
24	6.340	49	12.944	74	19.549	99	26.153
25	6.604	50	13.209	75	19.813	100	26.417

PVC (POLYVINYL CHLORIDE, TYPE I, GRADE 1, Meet the Requirements of ASTM D-1784)

This material is considered one of the most versatile and economical of the thermoplastics. It has a maximum service temperature of 140°F (60°C), and has been found satisfactory for handling most acids, bases, salts, and other corrosives. It may however, be damaged by ketones, a colorless, inflammable, volatile liquid, used as a paint remover and as a solvent for certain oils. PVC is used in process piping; that is, carrying liquids and gases, and for water service and industrial and laboratory chemical waste drainage.

CPVC (CHLORINATED POLYVINYL CHLORIDE, TYPE IV, GRADE 1, Meet the Requirements of ASTM D-1784)

This type of pipe is used for handling higher temperature corrosives, having a maximum service temperature of 210°F (99°C). It is comparable to PVC in its overall chemical resistance. Its uses include process piping for hot corrosive liquids, hot and cold water lines, and similar applications above the temperature range of PVC.

POLYPROPYLENE

This is the lightest thermoplastic piping material, but it has higher strength and better general chemical resistance than polyethylene and may be used at temperatures as high as 180°F (83°C). Polypropylene is an excellent material for laboratory and industrial drainage piping where mixtures of acids, bases, and solvents are involved. It has shown great resistance to sulfur-bearing compounds and is particularly useful in salt water disposal lines, low pressure gas gathering systems, and crude oil flow piping.

KYNAR (POLYVINYLIDENE FLUORIDE)

This is one of the fluoride containing thermoplastics for piping applications. It is particularly useful for industrial uses that involve chlorine solutions at temperatures up to 250°F (121°C). Other fluorine-containing thermoplastics include Teflon, Kel-f, and Halon. Because of their relatively high cost, these materials find usage as lining materials for metallic pipe.

POLYETHYLENE

This material has good chemical resistance and is generally satisfactory when used at temperatures below 120°F (49°C), but its mechanical strength is low. Types I and II (low and medium density) polyethylene are used frequently in chemical laboratory drainage lines, field irrigation, and for potable water systems.

ABS (ACRYLONITRITE BUTADIENE STYRENE)

This material has a high impact strength, is tough, and may be used at temperatures up to 180°F (82°C). It has lower chemical resistance and lower design strength than PVC; however, ABS is useful for carrying potable water, for irrigation, in gas lines, and in drain, waste, and vent piping. ABS may be joined by solvent welding or threading.

VITON

This material possesses exceptional resistance to a wide variety of chemicals. It is a fluorocarbon elastomer, a substance having rubber-like properties. Because of its inertness to chemicals, it is used extensively for O-ring seals in valves.

TEFLON

Virtually all ball valve seats and some other valve components are manufactured of TFE (Teflon). Because of its natural lubrication qualities, bearing surfaces of Teflon never need lubrication. Teflon is inert to virtually all chemicals.

SOLVENT WELDING INSTRUCTIONS FOR PVC AND CPVC PIPING JOINTS

Saw the pipe to desired length using a hand saw and miter box or a power saw. Remove all burrs with a knife, file, or sand paper.

Clean the connecting surfaces of both the pipe and fittings with a primer. With a brush of the correct size apply a complete coating of the primer material to the entire ID surface of the fitting socket and to an equivalent area on the OD of the pipe end. This will clean and etch the surfaces.

Apply solvent cement with a clean brush, first liberally to pipe OD, then lightly to fitting ID and again liberally to the pipe.

Cement should be applied to entire fitting socket including the shoulder at the socket bottom. All of pipe OD to be inserted in fitting must be covered with cement, including the cut end of the pipe.

Join the pipe and fitting immediately. Rotate fitting approximately ¼ turn as it is being pushed into pipe. The pipe should bottom in fitting. Hold the parts together until the cement takes hold to prevent pipe from backing out of the socket.

If the parts are correctly assembled, a bead of cement will appear between the pipe and the fitting. A lack of a bead indicates that an insufficient amount of cement was applied to the pipe.

PLASTIC THREADED JOINTS

Threading on plastic pipe reduces the effective wall thicknesses of the pipe and results in lower pressure ratings. Threaded connections should be used only with Schedule 80 or heavier pipe. Thread tape should be used for all threaded connections since threaded fittings tend to bind after long periods of service. Starting with the entry thread, wrap the tape tightly, covering all threads. Overlap each wrap ¼ in. Screw the fitting onto pipe and tighten it with strap wrenches. Avoid excessive torque; one to two turns past hand tight is adequate.

EXPANSION IN PLASTIC PIPING

The formula for calculating expansion or contraction in plastic piping is:

$$E = Y \frac{(T - F)}{10} \times \frac{L}{100}$$

Where E = expansion in inches
Y = constant factor expressing inches of expansion per 10° F temperature change per 100 feet of pipe.
T = maximum temperature
F = minimum temperature
L = length of pipe in running feet

EXAMPLE: How much expansion can be expected in 215 ft of PVC type I pipe installed at 75° F and operating at 135° F?

SOLUTION:

$$E = \frac{1}{3} \times \frac{60}{10} \times \frac{215}{100} = \frac{1}{3} \times 6 \times 2.15 = 4.3 \text{ in.}$$

Note: Allow for contraction when pipe is to be exposed to temperatures substantially below installation temperature.

Value of Y for Specific Plastics

Material	Y
PVC type I[a]	⅓
CPVC[b]	½
Polypropylene	½

[a]PVC = Polyvinyl chloride
[b]CPVC = Chlorinated polyvinyl chloride

Hot Water Demand per Fixture for Residences Gallons per Hour at 140°F (60°C)

Item	Gal/Hr at 140° F (60° C)
Lavatory (wash basin)	2
Bath tub	20
Dishwasher	15
Kitchen sink	10
Washing machine	20
Shower	75
Hourly heating capacity factor	30 percent
Storage capacity factor	70 percent

Total gph X hourly heating capacity factor = gph rating of tankless heater
Total gph X storage capacity factor = gallon capacity of storage tank

Capacity of Round Storage Tanks per Foot of Length

Diameter (in.)	Gal/Ft Length
18	13
20	16
24	24

DIMENSIONS OF COPPER AND BRASS PIPE

(Based on ASTM B251 by permission: American Society for Testing and Materials, Philadelphia, Pa.)

Copper Water Tube (ASTM B88)					Copper and Red Brass Pipe (ASTM B42 and B43)					
						Nominal Dimensions, in.				
		Nominal Wall Thickness, in.				Regular Weight			Extra Strong	
Standard Water Tube Size, in.	Actual Outside Diameter, in.	Type K*	Type L*	Type M*	Nominal Pipe Size in.	Outside Diameter	Inside Diameter	Wall Thickness	Inside Diameter	Wall Thickness
¼	0.375	0.035	0.030	. . .	⅛	0.405	0.281	0.062	0.205	0.100
⅜	0.500	0.049	0.035	. . .	¼	0.540	0.376	0.082	0.294	0.123
½	0.625	0.049	0.040	. . .	⅜	0.675	0.495	0.090	0.421	0.127
⅝	0.750	0.049	0.042	. . .	½	0.840	0.626	0.107	0.542	0.149
¾	0.875	0.065	0.045	. . .	¾	1.050	0.822	0.144	0.736	0.157
1	1.125	0.065	0.050	. . .	1	1.315	1.063	0.126	0.951	0.182
1¼	1.375	0.065	0.055	0.042	1¼	1.660	1.368	0.146	1.272	0.194
1½	1.625	0.072	0.060	0.049	1½	1.900	1.600	0.150	1.494	0.203
2	2.125	0.083	0.070	0.058	2	2.375	2.063	0.156	1.933	0.221
2½	2.625	0.095	0.080	0.065	2½	2.875	2.501	0.187	2.315	0.280
3	3.125	0.109	0.090	0.072	3	3.500	3.062	0.219	2.892	0.304
3½	3.625	0.120	0.100	0.083	3½	4.000	3.500	0.250	3.358	0.321
4	4.125	0.134	0.110	0.095	4	4.500	4.000	0.250	3.818	0.341
5	5.125	0.160	0.125	0.109	5	5.562	5.062	0.250	4.812	0.375
6	6.125	0.192	0.140	0.122	6	6.625	6.125	0.250	5.751	0.437
8	8.125	0.271	0.200	0.170	8	8.625	8.001	0.312	7.625	0.500
10	10.125	0.338	0.250	0.212	10	10.750	10.020	0.365	9.750	0.500
12	12.125	0.405	0.280	0.254	12	12.750	12.000	0.375		

* Recommendations:

Type K: General Plumbing and heating systems and underground service, for severe conditions.

Type L: For interior use in general plumbing and heating

Type M: Non-pressure applications (drain, vents, etc.)

DIMENSIONS OF WELDED AND SEAMLESS STEEL PIPE (ASA B36.10)

(Listed as Standard Wall, Extra Strong Wall and Double Extra Strong Wall)

Nominal Pipe Size	Outside Diameter	Nominal Wall Thickness		
		Standard Wall	Extra Strong Wall	Double Extra Strong Wall
1/8	0.405	**0.068**	**0.095**	....
1/4	0.540	**0.088**	**0.095**	....
3/8	0.675	**0.091**	**0.126**	....
1/2	0.840	**0.109**	**0.147**	0.294
3/4	1.050	**0.113**	**0.154**	0.308
1	1.315	**0.133**	**0.179**	0.358
1 1/4	1.660	**0.140**	**0.191**	0.382
1 1/2	1.900	**0.145**	**0.200**	0.400
2	2.375	**0.154**	**0.218**	0.436
2 1/2	2.875	**0.203**	**0.276**	0.552
3	3.500	**0.216**	**0.300**	0.600
3 1/2	4.000	**0.226**	**0.318**	....
4	4.500	**0.237**	**0.337**	0.674
5	5.563	**0.258**	**0.375**	0.750
6	6.625	**0.280**	**0.432**	0.864
8	8.625	**0.322**	**0.500**	0.875
10	10.750	**0.365**	**0.500**	....
12	12.750	0.375	0.500	....
14	14.000	0.375	0.500	....
16	16.000	0.375	0.500	....
18	18.000	0.375	0.500	....
20	20.000	0.375	0.500	....
24	24.000	0.375	0.500	....

All dimensions given in inches.

The decimal thicknesses listed for the respective pipe sizes represent their nominal or average wall dimensions. For tolerances on wall thicknesses, see appropriate material specifications.

Thicknesses shown in bold face type for Standard Wall are identical with corresponding thicknesses shown in bold face type for Schedule 40 in Table 3.18. Those shown in bold face type for Extra Strong Wall are identical with corresponding thicknesses shown in bold face type in Schedules 60 and 80 in Table 3.18.

Double Extra Strong Wall has no corresponding schedule numbers.

DIMENSIONS OF WELDED AND SEAMLESS STAINLESS STEEL PIPE

Nominal Pipe Size	Outside Diameter	Nominal Wall Thickness			
		Schedule 5S†	Schedule 10S†	Schedule 40S	Schedule 80S
1/8	0.405	. . .	0.049	0.068	0.095
1/4	0.540	. . .	0.065	0.088	0.119
3/8	0.675	. . .	0.065	0.091	0.126
1/2	0.840	0.065	0.083	0.109	0.147
3/4	1.050	0.065	0.083	0.113	0.154
1	1.315	0.065	0.109	0.133	0.179
1 1/4	1.660	0.065	0.109	0.140	0.191
1 1/2	1.900	0.065	0.109	0.145	0.200
2	2.375	0.065	0.109	0.154	0.218
2 1/2	2.875	0.083	0.120	0.203	0.276
3	3.500	0.083	0.120	0.216	0.300
3 1/2	4.000	0.083	0.120	0.226	0.318
4	4.500	0.083	0.120	0.237	0.337
5	5.563	0.109	0.134	0.258	0.375
6	6.625	0.109	0.134	0.280	0.432
8	8.625	0.109	0.148	0.322	0.500
10	10.750	0.134	0.165	0.365	0.500*
12	12.750	0.156	0.180	0.375*	0.500*

All dimensions are given in inches.

The decimal thicknesses listed for the respective pipe sizes represent their normal or average wall dimensions.

Tolerances: +12.5%.

* These do not conform to ASA B36.10 Schedule Numbers, but correspond to standard weight (0.375) and extra strong (0.500).

† Schedule 5S and 10S wall thicknesses do not permit threading in accordance with ASA B2.1.

Steel Butt-Welding Fittings for Use with Standard Pipe Dimensions in Inches

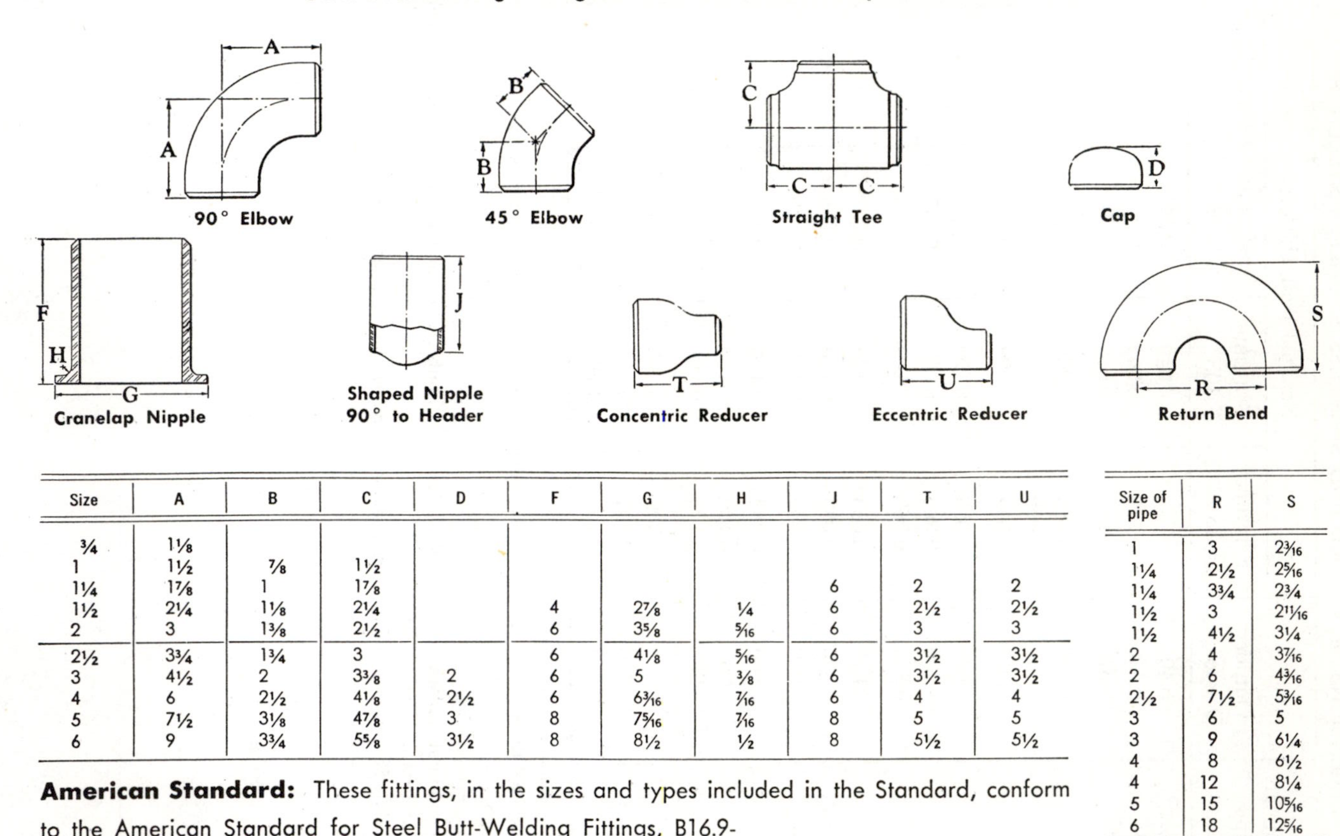

Size	A	B	C	D	F	G	H	J	T	U
¾	1⅛									
1	1½	⅞	1½							
1¼	1⅞	1	1⅞					6	2	2
1½	2¼	1⅛	2¼		4	2⅞	¼	6	2½	2½
2	3	1⅜	2½		6	3⅝	5/16	6	3	3
2½	3¾	1¾	3		6	4⅛	5/16	6	3½	3½
3	4½	2	3⅜	2	6	5	⅜	6	3½	3½
4	6	2½	4⅛	2½	6	6 3/16	7/16	6	4	4
5	7½	3⅛	4⅞	3	8	7 5/16	7/16	8	5	5
6	9	3¾	5⅝	3½	8	8½	½	8	5½	5½

Size of pipe	R	S
1	3	2 3/16
1¼	2½	2 5/16
1¼	3¾	2¾
1½	3	2 11/16
1½	4½	3¼
2	4	3 7/16
2	6	4 3/16
2½	7½	5 3/16
3	6	5
3	9	6¼
4	8	6½
4	12	8¼
5	15	10 5/16
6	18	12 5/16

American Standard: These fittings, in the sizes and types included in the Standard, conform to the American Standard for Steel Butt-Welding Fittings, B16.9-

Standard Cast-Iron Fittings—Dimensions in Inches

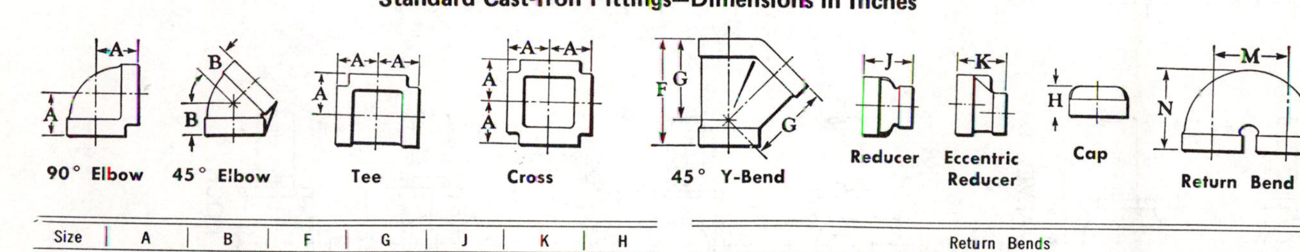

90° Elbow 45° Elbow Tee Cross 45° Y-Bend Reducer Eccentric Reducer Cap Return Bend

Size	A	B	F	G	J	K	H
1/4	13/16	3/4					
3/8	15/16	13/16					
1/2	1 1/8	7/8					
3/4	1 5/16	1	3	2 1/4	1 1/2		
1	1 1/2	1 1/8	3 1/2	2 3/4	*1 7/8		
1 1/4	1 3/4	1 5/16	4 1/4	3 1/4	2 1/8	2 1/8	
1 1/2	1 15/16	1 7/16	4 7/8	3 13/16	2 1/4	2 1/4	
2	2 1/4	1 11/16	5 3/4	4 1/2	2 7/16	2 7/16	
2 1/2	2 11/16	1 15/16	6 3/4	5 3/16	2 5/8	2 11/16	
3	3 1/8	2 3/16	7 7/8	6 1/8	2 7/8	2 15/16	
3 1/2	3 7/16	2 3/8				3 1/8	
4	3 3/4	2 5/8	9 3/4	7 5/8	3 3/8	3 3/8	2 1/16
5	4 1/2	3 1/16			3 7/8	3 7/8	2 3/8
6	5 1/8	3 7/16			4 3/8	4 3/8	2 5/8
8	6 9/16	4 1/4			5 1/4	5 1/4	3 1/8

Return Bends								
Size	M	N	Size	M	N	Size	M	N
Close Pattern			Open Pattern			Wide Pattern		
1/2	1 1/4	1 23/32	3/4	1 7/8	2 7/32	1	3	3
3/4	1 1/2	2 1/32	1	2 1/2	2 11/16	1	4	3 1/2
1	1 3/4	2 3/8	1 1/4	3	3 9/32	1 1/4	4	3 3/4
1 1/4	2 1/4	2 29/32	1 1/2	3 1/2	3 3/4	1 1/4	6	4 3/4
1 1/2	2 1/2	3 1/4	2	4 1/2	4 19/32	1 1/2	6	5
2	3 1/4	3 31/32	2 1/2	5 1/2	5 7/16	2	6	5 5/16
			3	6 1/2	6 5/16			

Close Pattern Return Bends can not be used to make up parallel coils. The center to center dimension is so close that the bands of adjacent bends will not clear each other.

* 1x1/2-inch Reducers are 1 11/16 inches end to end.

Screwed Valves

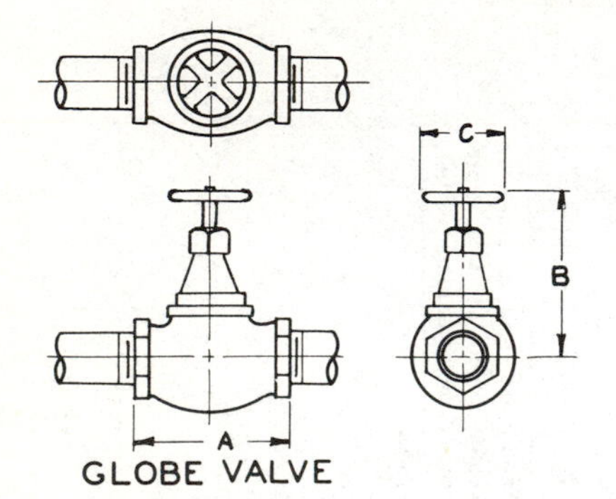

GLOBE VALVE

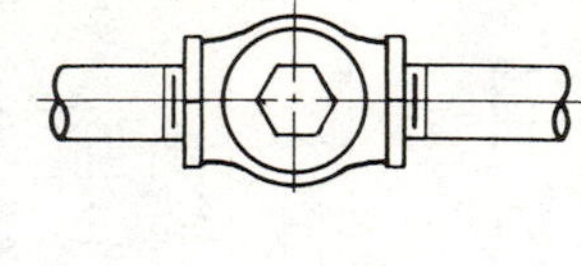
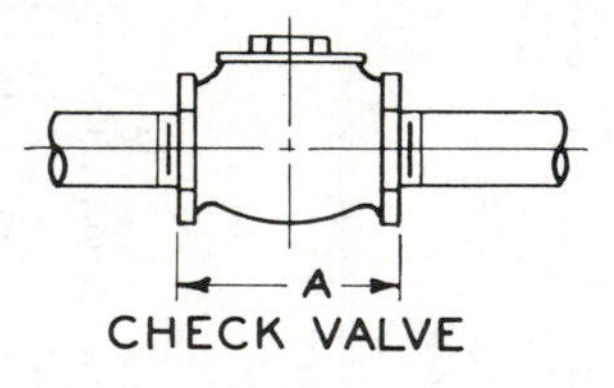

CHECK VALVE

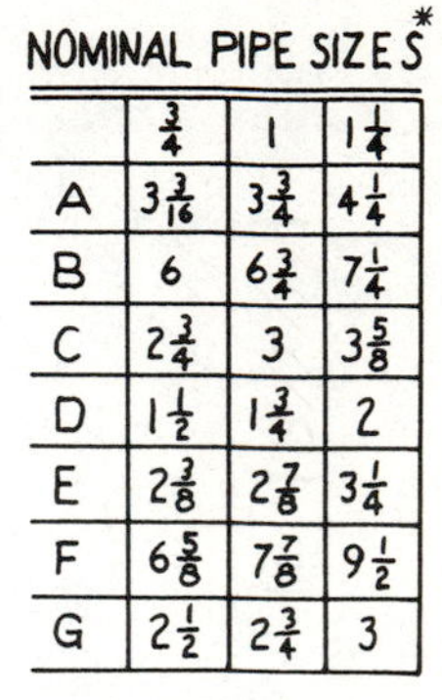

NOMINAL PIPE SIZES*

	$\frac{3}{4}$	1	$1\frac{1}{4}$
A	$3\frac{3}{16}$	$3\frac{3}{4}$	$4\frac{1}{4}$
B	6	$6\frac{3}{4}$	$7\frac{1}{4}$
C	$2\frac{3}{4}$	3	$3\frac{5}{8}$
D	$1\frac{1}{2}$	$1\frac{3}{4}$	2
E	$2\frac{3}{8}$	$2\frac{7}{8}$	$3\frac{1}{4}$
F	$6\frac{5}{8}$	$7\frac{7}{8}$	$9\frac{1}{2}$
G	$2\frac{1}{2}$	$2\frac{3}{4}$	3

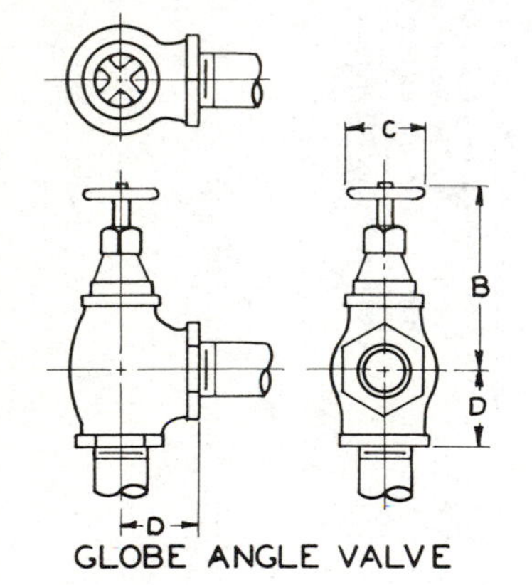

GLOBE ANGLE VALVE

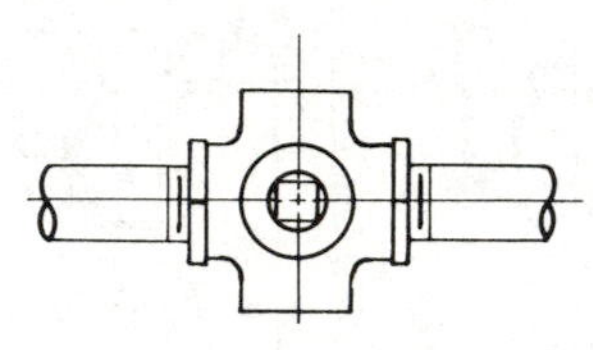
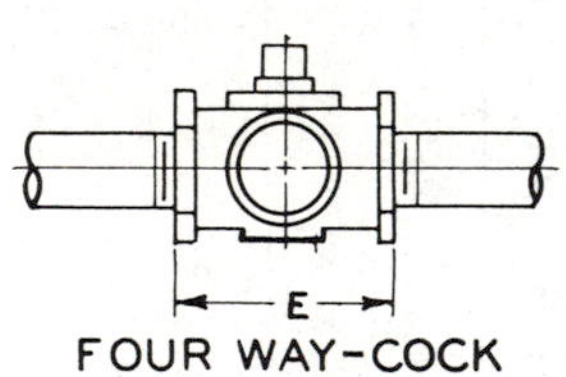

FOUR WAY-COCK

	$1\frac{1}{2}$	2	$2\frac{1}{2}$
A	$4\frac{3}{4}$	$5\frac{3}{4}$	$6\frac{3}{4}$
B	$8\frac{1}{4}$	$9\frac{1}{2}$	11
C	4	$4\frac{3}{4}$	6
D	$2\frac{1}{4}$	$2\frac{3}{4}$	$3\frac{1}{4}$
E	$3\frac{1}{2}$	$3\frac{7}{8}$	$4\frac{1}{2}$
F	$10\frac{7}{8}$	$13\frac{1}{8}$	$15\frac{3}{8}$
G	$3\frac{5}{8}$	4	$4\frac{3}{8}$

* *Dimensions compiled from manufacturers' catalogues for drawing purposes.*

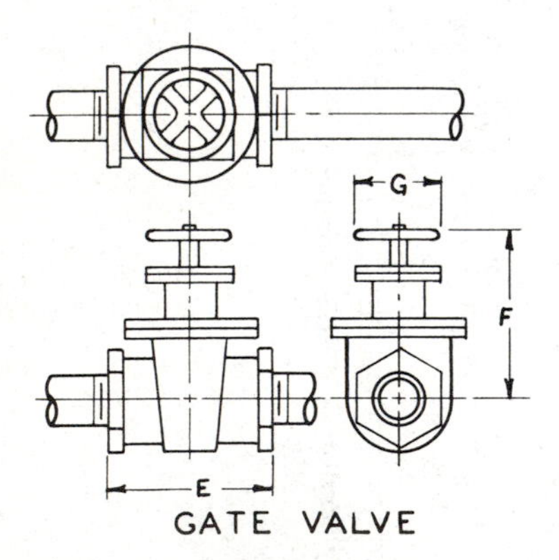

GATE VALVE

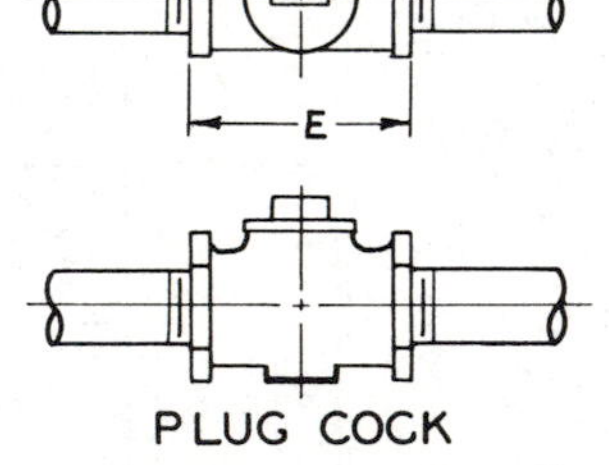

PLUG COCK

American Standard Taper Pipe Threads

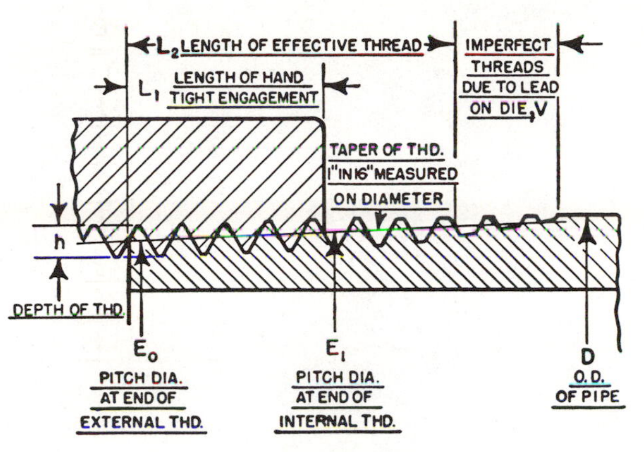

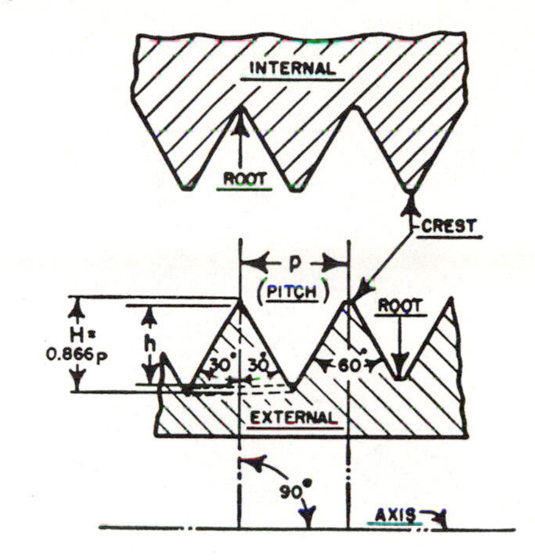

$E_0 = D - (0.050D + 1.1)p$
$E_1 = E_0 + 0.0625\ L_1$
$L_2 = (0.80D + 6.8)p$
$H = 0.866p$

Nominal Pipe Size Inches	Dimensions, Inches								Pitch of Thread
	D	No. Threads per inch N	E_0	E_1	L_2	L_1	V	h	p
1/8	.405	27	.36351	.37476	.2639	.180	0.1285	.02963	.03704
1/4	.540	18	.47739	.48989	.4018	.200	0.1928	.04444	.05556
3/8	.675	18	.61201	.62701	.4078	.240	0.1928	.04444	.05556
1/2	.840	14	.75843	.77843	.5337	.320	0.2478	.05714	.07143
3/4	1.050	14	.96768	.98887	.5457	.339	0.2478	.05714	.07143
1	1.315	11½	1.21363	1.23863	.6828	.400	0.3017	.06957	.08696
1¼	1.660	11½	1.55713	1.58338	.7068	.420	0.3017	.06957	.08696
1½	1.900	11½	1.79609	1.82234	.7235	.420	0.3017	06957	.08696
2	2.375	11½	2.26902	2.29627	.7565	.436	0.3017	.06957	.08696
2½	2.875	8	2.71953	2.76216	1.1375	.682	0.4337	.10000	.12500
3	3.500	8	3.34062	3.38850	1.2000	.766	0.4337	.10000	.12500
3½	4.000	8	3.83750	3.88881	1.2500	.821	0.4337	.10000	.12500
4	4.500	8	4.33438	4.38712	1.3000	.844	0.4337	.10000	.12500
5	5.563	8	5.39073	5.44929	1.4063	.937	0.4337	.10000	12500
6	6.625	8	6.44609	6.50597	1.5125	.958	0.4337	.10000	.12500
8	8.625	8	8.43359	8.50003	1.7125	1.063	0.4337	.10000	.12500
10	10.750	8	10.54531	10.62094	1.9250	1.210	0.4337	.10000	.12500
12	12.750	8	12.53281	12.61781	2.1250	1.360	0.4337	.10000	.12500
14	14.000	8	13.77500	13.87262	2.2500	1.562	0.4337	.10000	.12500
16	16.000	8	15.76250	15.87575	2.4500	1.812	0.4337	.10000	.12500
18	18.000	8	17.75000	17.87500	2.6500	2.000	0.4337	.10000	.12500
20	20.000	8	19.73750	19.87031	2.8500	2.125	0.4337	.10000	.12500
24	24.000	8	23.71250	23.86094	3.2500	2.375	0.4337	.10000	.12500

Data from Pipe Threads, ASA B2.1 American Society of Mechanical Engineers.

Index